Statistical Graphics Procedures by Example
Effective Graphs Using SAS®

Sanjay Matange
Dan Heath

The correct bibliographic citation for this manual is as follows: Matange, Sanjay, and Dan Heath. 2011. *Statistical Graphics Procedures by Example: Effective Graphs Using SAS®*. Cary, NC: SAS Institute Inc.

Statistical Graphics Procedures by Example: Effective Graphs Using SAS®

Copyright © 2011, SAS Institute Inc., Cary, NC, USA

ISBN 978-1-60764-887-1 (electronic book)
ISBN 978-1-60764-762-1

All rights reserved. Produced in the United States of America.

For a hard-copy book: No part of this publication may be reproduced, stored in a retrieval system, or transmitted, in any form or by any means, electronic, mechanical, photocopying, or otherwise, without the prior written permission of the publisher, SAS Institute Inc.

For a Web download or e-book: Your use of this publication shall be governed by the terms established by the vendor at the time you acquire this publication.

The scanning, uploading, and distribution of this book via the Internet or any other means without the permission of the publisher is illegal and punishable by law. Please purchase only authorized electronic editions and do not participate in or encourage electronic piracy of copyrighted materials. Your support of others' rights is appreciated.

U.S. Government Restricted Rights Notice: Use, duplication, or disclosure of this software and related documentation by the U.S. government is subject to the Agreement with SAS Institute and the restrictions set forth in FAR 52.227-19, Commercial Computer Software-Restricted Rights (June 1987).

SAS Institute Inc., SAS Campus Drive, Cary, North Carolina 27513-2414

1st printing, November 2011

SAS® Publishing provides a complete selection of books and electronic products to help customers use SAS software to its fullest potential. For more information about our e-books, e-learning products, CDs, and hard-copy books, visit the SAS Publishing Web site at **support.sas.com/publishing** or call 1-800-727-3228.

SAS® and all other SAS Institute Inc. product or service names are registered trademarks or trademarks of SAS Institute Inc. in the USA and other countries. ® indicates USA registration.

Other brand and product names are registered trademarks or trademarks of their respective companies.

Contents

Preface   ix

Chapter 1   Introduction
- 1.1   Principles of Effective Graphics   3
- 1.2   Automatic Graphs from SAS Procedures   10
- 1.3   Graph Template Language   11
- 1.4   Statistical Graphics Procedures   11
- 1.5   Organization of This Book   12
- 1.6   Data Sets and Custom Styles   14
- 1.7   Color and Gray-Scale Graphs   14
- 1.8   Effective Graphics and the Use of Decorative Skins   15
- 1.9   SAS 9.2 and SAS 9.3 Features   15

Chapter 2   Statistical Graphics Procedures
- 2.1   Key Concepts   19
- 2.2   SGPLOT Procedure   22
- 2.3   SGPANEL Procedure   24
- 2.4   Combining Statements   29
- 2.5   SGSCATTER Procedure   30
- 2.6   Styles and Their Usage   32
- 2.7   Template-Based Graphics vs. Device-Based Graphics   33

Chapter 3   Common Graphs
- 3.1   Introduction   37
- 3.2   Single-Cell Graphs   38
- 3.3   Classification Panels   55
- 3.4   Comparative and Matrix Graphs   57

Chapter 4   Basic Plots
- 4.1   Introduction   61
- 4.2   SGPLOT Procedure   62
- 4.3   Plot Roles and Options   63
- 4.4   Scatter Plot   64
- 4.5   Scatter Plots with Data Labels   74

- 4.6 Series Plot 78
- 4.7 Step Plot 84
- 4.8 Band Plot 89
- 4.9 Needle Plot 94
- 4.10 Vector Plot 97
- 4.11 VBarParm and HBarParm Plots (9.3) 100
- 4.12 Bubble Plot (9.3) 105
- 4.13 HighLow Plot (9.3) 108
- 4.14 Reference Lines 112
- 4.15 Parametric Line Plot (9.3) 115
- 4.16 Waterfall Chart (9.3) 117
- 4.17 Combining the Plots 120

Chapter 5 Fit and Confidence Plots
- 5.1 Introduction 129
- 5.2 Fit Plot Roles and Options 130
- 5.3 Regression Plot 131
- 5.4 Loess Plot 138
- 5.5 Penalized B-Spline Plot 145
- 5.6 Ellipse Plot 152
- 5.7 Combining the Plots 157

Chapter 6 Distribution Plots
- 6.1 Introduction 161
- 6.2 Histogram 162
- 6.3 Density Plot 166
- 6.4 Vertical Box Plot 169
- 6.5 Horizontal Box Plot 174
- 6.6 Combining the Plots 177

Chapter 7 Categorization Plots
- 7.1 Introduction 181
- 7.2 Categorization Plot Roles and Common Options 182
- 7.3 Vertical Bar Charts 183
- 7.4 Horizontal Bar Charts 191
- 7.5 Vertical Line Charts 198
- 7.6 Horizontal Line Charts 205

7.7   Dot Plots   212
  7.8   Combining the Plots   218

Chapter 8 Axes, Legends, and Insets
  8.1   Introduction   221
  8.2   Linear Axis   223
  8.3   Log Axis   225
  8.4   Time Axis   227
  8.5   Discrete Axis   229
  8.6   Legends   230
  8.7   Insets   232

Chapter 9 Annotation and Attribute Maps (SAS 9.3)
  9.1   Annotation   235
  9.2   Attribute Maps (9.3)   251

Chapter 10 Classification Panels
  10.1   Introduction   257
  10.2   SGPANEL Procedure   258
  10.3   PANELBY Statement   259
  10.4   Classification Panels   261
  10.5   Paging of Large Panels   266

Chapter 11 Comparative and Matrix Plots
  11.1   Introduction   269
  11.2   SGSCATTER Procedure   269
  11.3   PLOT Statement   270
  11.4   COMPARE Statement   276
  11.5   MATRIX Statement   281

Chapter 12 Health and Life Sciences Graphs
  12.1   Introduction   287
  12.2   Forest Plot   288
  12.3   Forest Plot (9.3)   289
  12.4   Survival Plot   290
  12.5   Adverse Event Timeline   291
  12.6   Adverse Event Timeline (9.3)   292

12.7 Maximum LFT Values by Treatment (9.3)   293
12.8 Median of Lipid Profile over Time by Treatment (9.3)   294
12.9 QTc Change from Baseline over Time by Treatment (9.3)   295
12.10 QTc Change Graph with Annotated "At Risk" Values (9.3)   296
12.11 QTc Change from Baseline over Time by Treatment   297
12.12 LFT Safety Panel, Baseline vs. Study   298
12.13 Immunology Profile by Treatment   299
12.14 Most Frequent On-Therapy Adverse Events by Frequency   300
12.15 LFT Patient Profile   301
12.16 Panel of LFT Values   302
12.17 Distribution of Eye Irritation Using PROC SGPANEL (9.3)   303
12.18 Distribution of Eye Irritation Using PROC SGPLOT (9.3)   304
12.19 Vital Signs by Time Point Name   305
12.20 Concomitant Medications   306
12.21 Creating a 2 x 2 Cell Graph Using PROC SGPLOT   307

Chapter 13   Business Graphs

13.1 Introduction   311
13.2 Stock Price and Volume Chart   312
13.3 Financial Trend and Bond Maturity Graph (9.3)   313
13.4 Danger of High P/E Ratios   314
13.5 Oil Consumption Trend by Country   315
13.6 Product Sales and Target Graph (9.3)   316
13.7 Social Network (9.3)   317

Chapter 14   Styles

14.1 Introduction   321
14.2 Using Styles   321
14.3 Style Elements   321
14.4 Using Style Elements   324
14.5 Style Element Usage Precedence   324

Chapter 15   ODS Destination and ODS Graphics Options

15.1 Introduction   327
15.2 ODS Destination Options   327
15.3 ODS Graphics Options   328

Chapter 16 Tips for Graph Output

- 16.1 Introduction 333
- 16.2 Creating Small Graphs for Use in Documents 333
- 16.3 Creating Large Graphs for Use in Presentations 334
- 16.4 Combining Graph Size and DPI 336
- 16.5 Impact of Graph Size on System Resources 336

Preface

As we looked with anticipation to the first release of the Statistical Graphics Procedures in SAS 9.2, it became evident to us that these procedures truly represented a new way of creating analytical graphs. We started thinking about the best ways to communicate the features and capabilities of these new procedures. While we discussed different approaches, we found ourselves receiving a lot of questions from users about the procedures. Often our answers to them were in the form of examples. As we created more examples in such areas as clinical safety, pharmaceuticals, health care, and finance, we soon realized that the best way to communicate the features of these procedures was through such examples.

The question we often got from users was simply, "How do I make this graph?" The best way for us to answer that question was to show the code needed to create it. If a book could answer that question, what would it look like? Why not create a book that starts from the end result and work backward to show exactly what is needed to create the graph? That led to the idea of writing this book.

The primary audience for this book is the SAS user who wants to visualize raw data or create a graph from the results of a custom analysis. Often, you already have a mental image of the graph you want to create; you just need to quickly find the correct syntax. Product documentation and books on the topic often take a procedure-centric approach and describe the features one at a time. Figuring out what is needed to create a graph from such resources requires a solid understanding of the procedure, and it can take a while to obtain the right results. It would be much easier if we start with the graph you want, and show you the code instead.

This book addresses this situation by using examples to document the procedure options. Users can look through the large number of graph examples and find the type of graph they want to create. Each of the graph examples includes the code needed to generate the graph, along with a brief commentary on key features of the graph.

The reason why this approach works so well for the SG Procedures is because these procedures take a building-block approach. You start with the basic plot, and simply add the features you need one at a time. The procedures support a wide array of plot types, so the combinations and possibilities grow rapidly. For example, if you know how to build a simple series plot, then creating a plot with three series plots becomes straightforward. You simply add two more series statements, and the procedure automatically does the work. The same principle also applies to combinations of disparate, but compatible, plot types, such as a bar chart and a line chart.

The book also describes, by example, other important features of the procedures such as axes, insets and legends. For example, you can take a sample graph with a linear axis, and change it to log. You can also create a custom axis displaying only the values you want, add insets and customize the legends.

For an overview of the organization of this book, see section 1.5.

The SG Procedures are designed with the principles of effective graphics built-in to convey the information with maximum clarity and minimum clutter. By default, these procedures will create graphs that are free of unnecessary clutter in the graph elements, legends, and axes.

These procedures are designed to create graphs that are suitable for the statistical and analytical use cases. Such graphs emphasize the maximization of "chart ink" and removal of elements that are not clearly necessary for the delivery of the information. It will be evident that you have to do very little to get aesthetically pleasing graphs for these use cases.

Visual aesthetics, however, are in the eye of the beholder. In non-statistical use case, the expectation of the consumer is for flashy graphs, even at the cost of some effectiveness. One person's "chart junk" is another person's "cool". These procedures are finding increasing usage for the creation of graphs for the business domain. They support some options to add "flash" to these graphs. These options are mainly available for the bar charts and can be used when necessary.

SG Procedures support full-color graphics. Many styles shipped with SAS are optimized for creation of color graphs. However, printing color graphs in gray scale can sometimes lead to undesirable results. This is important, since many technical journals are printed in gray scale. Since this book is printed in gray scale we have used the appropriate gray scale styles.

The examples and techniques discussed in this book will be relevant and useful for all SAS users. This is particularly so for statisticians and other analytical users in the pharmaceutical, clinical trials, health care, financial, and other domains. This book is focused on how to create the required graph given the data. Techniques for modeling and analysis of the data itself are beyond the scope of this book.

Acknowledgements

Many people have contributed in many different ways to make this book possible. We would like to thank Bob Rodriguez, Senior Director in Advanced Analytics at SAS, for supporting the concept behind this book and for his detailed review of the contents. His insightful suggestions have significantly improved the quality of the materials and the presentation of the book.

On the contents of the book and accuracy of the information, we received invaluable support from Melisa Turner and Susan Schwartz, our "eagle eye" team of technical reviewers. Both Melisa and Susan invested many days reviewing the code and improving the contents. We also thank our reviewers Lelia McConnell, David Schlotzhauer, Peter Christie, and Rebecca Ottesen for their valuable suggestions. Our heartfelt thanks go to Susan Slaughter for her review, moral support, and guidance on this project. The table of Statement Combinations in section 2.4 is modeled after a similar table from Susan and Lora Delwiche's recent paper, "Using PROC SGPLOT for Quick High-Quality Graphs."[i]

Last and most importantly, we thank our families for the understanding and support they provided while we spent long evenings and weekends on this project.

Author Pages

Each SAS Press author has an author page, which includes several features that relate to the author including a biography, book descriptions for coming soon titles and other titles by the author, contact information, links to sample chapters and example code and data, events and extras, and more.

You can access our author pages at http://support.sas.com/publishing/authors/matange.html and http://support.sas.com/publishing/authors/heath.html.

Comments or Questions?

If you have comments or questions about this book, you may contact the authors through SAS as follows:

Mail:

SAS Institute Inc.
SAS Press
Attn: <Author's name>
SAS Campus Drive
Cary, NC 27513

E-mail: saspress@sas.com

Fax: (919) 677-4444

Please include the title of the book in your correspondence.

For a complete list of books available through SAS Press, visit support.sas.com/publishing.

SAS Publishing News:

Receive up-to-date information about all new SAS publications via e-mail by subscribing to the SAS Publishing News monthly eNewsletter. Visit support.sas.com/subscribe.

---

[i] Available at http://support.sas.com/resources/papers/proceedings09/158-2009.pdf.

Chapter 1

Introduction

1.1 Principles of Effective Graphics   3
1.2 Automatic Graphs from SAS Procedures   10
1.3 Graph Template Language   11
1.4 Statistical Graphics Procedures   11
1.5 Organization of This Book   12
1.6 Data Sets and Custom Styles   14
1.7 Color and Gray-Scale Graphs   14
1.8 Effective Graphics and the Use of Decorative Skins   15
1.9 SAS 9.2 and SAS 9.3 Features   15

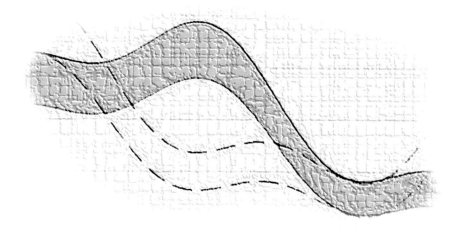

"Then there is the man who drowned crossing a stream
with an average depth of six inches."
~W.I.E. Gates

Chapter 1: Introduction

Graphs are an essential part of modern data analysis. From clinical trials to quality control, effective graphs are integral to the analysis process. Large quantities of data are collected for clinical drug trials for safety, retail sales, warranty claims, medical lab results, and financial transactions. Analysis of this data often relies on review of the data in tabular form. Viewing the data in the form of a graph along with results of the statistical analysis of the data on the same graph can significantly enhance the understanding of the data and the results.

A key aspect of this process is the ability to create an effective graph that can communicate the raw data along with the statistical analysis results in a clear and concise form. These graphs can help the analyst to visualize the trends and patterns in the data and the associations between variables that are not evident in tabular form. Such insights can guide the direction of further questions and formulation of additional testing methods and gathering of more focused data.

1.1 Principles of Effective Graphics

Research in the field of visual perception has guided the formulation of the principles for creation of effective graphs. *The Visual Display of Quantitative Information* (2001) by Edward Tufte, *Visualizing Data* (1993) by William S. Cleveland, and *Creating More Effective Graphs* (2004) by Naomi Robbins provide an in-depth coverage of the topic. Some of these principles are reviewed here.

A graph is considered effective if it conveys the intended information in a way that can be understood quickly and without ambiguity by most consumers. Figure 1.1 shows an interesting graph of the levels of usage of text messaging on cell phones by teenagers. The font sizes are increased for readability.

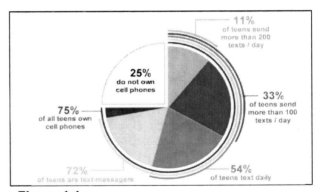

Figure 1.1

In Figure 1.1, the consumer would be hard pressed to discern the differences (if any) between individual pie slices such as the second and third slice clockwise from the top. The pie chart does not facilitate such comparison of magnitude between individual categories.

This graph is further complicated by addition of the cumulative strips along the outside of the pie chart. One reason why a pie chart is not an effective graph is the difficulty of making magnitude comparisons when the data is plotted as an angle from a common (or non-common) base. This applies to the strips and the slices.

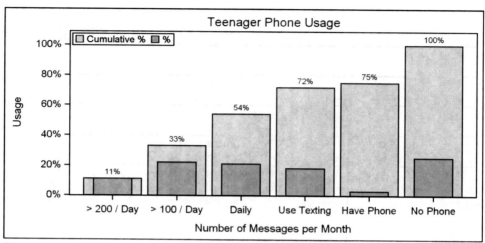

Figure 1.2

Comparison of magnitude along a linear scale from a common base is very reliable. The same data rendered as a bar chart is much easier to decode as shown above in Figure 1.2. In this graph, the comparisons between the 2nd, 3rd and 4th bars for the % case are much easier and more reliable.

The pie chart can be a useful visual for some use cases as shown in Figure 1.3. This graph shows the portion of sales for Auto as a fraction of the total sales. The pie chart can work well for visualizing such "part-to-whole" comparisons.

1.1.1 Visual Perception

The concepts of attentive and pre-attentive vision are summarized by Daniel Carr in *Information Visualization: Perception for Design* (2004) by Colin Ware. Attentive vision requires scrutiny of the object and an active participation on the part of the observer.

On the other hand, pre-attentive vision allows processing to be done prior to conscious attention. Usage of pre-attentive features helps in rapid discrimination between various artifacts of a graph.

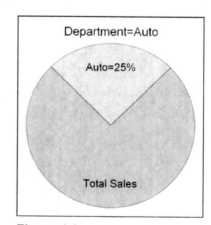

Figure 1.3

Differentiation between length of lines is pre-attentive regardless of a common baseline.

In Figure 1.4, it is relatively easy to compare the magnitudes of the responses for all drugs. The line segments in Figure 1.5 are plotted from different baselines, but still it is possible to compare the magnitudes of each line segment for the drugs.

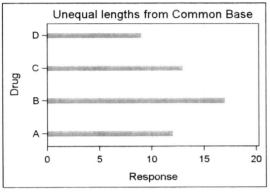
Figure 1.4

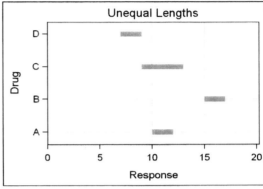
Figure 1.5

Differentiation of marker shapes and line patterns is pre-attentive. Groups can be easily differentiated when marker shapes or line patterns are used as grouping indicators as shown in Figures 1.6 and 1.7.

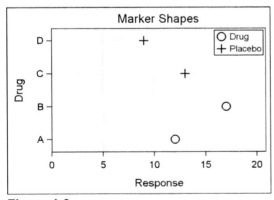
Figure 1.6

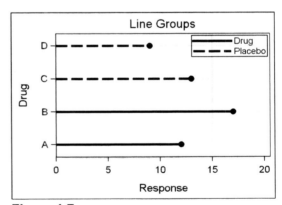
Figure 1.7

Based on their pre-attentive nature (ease of decoding) we conclude the following:

- Distances from a common baseline are useful for magnitude comparisons.
- Marker shape is effective for classification.
- Line pattern is effective for classification.
- Color is also an effective tool for classification, but since this book is published in gray scale, we have not utilized that feature.

6 *Statistical Graphics Procedures by Example: Effective Graphs Using SAS*

1.1.2 Accuracy of Magnitude Perception

Stevens' power law proposes a relationship between the magnitude of a physical stimulus and its perceived intensity or strength. The law proposes that the accuracy of magnitude perception for visual length is linear. Magnitude of a line twice as long is perceived almost as twice as long. However, the accuracy of perception of magnitude is reduced for other representations. For area, it is only about 1.6. That is, an area twice as large only seems like 1.6 times as large. So, we tend to underestimate areas.

For representation of magnitude, we can conclude the following:

- Linear distance on a common scale and base line is most effective.
- Linear distance on a common but non-aligned scale is effective.
- Areas are only about 70% as effective as linear distance.
- Volumes are only about 60% as effective as linear distance.
- Angular distance from a common base is not very effective.
- Angular distance from non-common base is ineffective.
- Color in general is not an effective representation of magnitude.
- Color intensity can be used as a relative measure of magnitude.
- Color hue is not a good representation of magnitude.
- 3-D plots distort the perception of absolute magnitude.

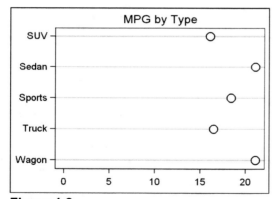

Figure 1.8   **Figure 1.9**

Figure 1.8 and Figure 1.9 both display the mean MPG by type of car. Figure 1.8 uses the dot plot where the response values are plotted as a linear distance from a common baseline. It is very easy for the eye to decode the relative values of each car type. Grid lines help to line up the values.

Figure 1.9 displays the same data using a pie chart. Clearly, it is much harder to decode the relative values since it is difficult to make good magnitude comparisons using angular distances, especially from different baselines.

1.1.3 Usage of 3-D Graphs

Figure 1.10 and Figure 1.11 both display the mean MPG by type of car. Figure 1.10 uses a vertical bar chart to display the data. It is very easy for the eye to decode the relative values for each car type. A format is applied to the response column to reduce the clutter for the data label.

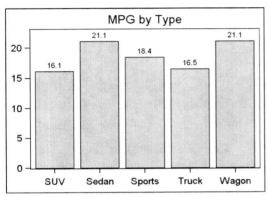

Figure 1.10

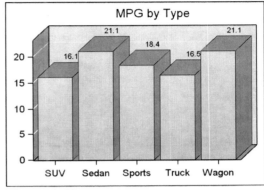

Figure 1.11

Figure 1.11 displays the same data using extruded 3-D bars. This is often referred to as a 2.5-D graph, since the data itself has only two dimensions. The third dimension is artificially added to make the bars appear like 3-D blocks. Usage of such aesthetic features can sometimes inhibit the process of decoding the data accurately.

There are several potential pitfalls in the 2.5-D representation of the data. The axis values are displayed on the left at the "front" face of the bars. The grid lines are drawn along the side and back face of the graph. So to measure the value for each bar, one has to line up the correct face (front or back).

Often in such representations, the bars do not occupy the full depth of the walls, thus leaving room for confusion. Even though the bar values are displayed in the 2.5-D case, some values, such as for the Truck category, can become partially hidden behind other bars.

In general the following conclusions can be drawn:

- o Dot plots, needle plots, and bar charts are good for representations of magnitude.
- o Pie charts and area plots are not ideal for representations of magnitude.
- o Color intensity is not an effective representation of magnitude.
- o 3-D representations are often not effective when the data are 2-D.
- o Unobtrusive grid lines can help in the decoding of the data.

1.1.4 Proximity Increases the Accuracy of Comparisons

When you compare magnitude between categories, closer proximity increases the accuracy of comparison. For an effective graph, it helps to bring the items that are to be compared as close to each other as possible.

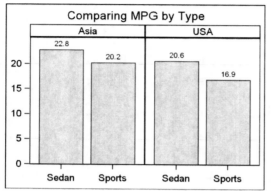
Figure 1.12

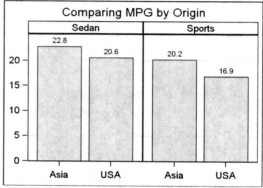
Figure 1.13

Figure 1.12 is suitable for comparison of MPG for sedans and sports cars manufactured in different regions (origin). The comparison between car types is facilitated by bringing these categories closer in proximity. In this graph it is harder to compare "Sedan" from USA with "Sedan" from Asia. Figure 1.13 is more suitable for such a comparison where "Origin" is used as the category role.

1.1.5 Simplify and Reduce Clutter

Edward Tufte's principles for the creation of effective graphics include the following recommendations:

1. Eliminate "chart junk."
2. Maximize "data ink."

Often when creating graphics for marketing and sales presentations, there is a desire to make the graph visually "compelling". To add this "Wow" factor, visual elements may be added to the graph to make it more aesthetically appealing. If one is not careful, these artifacts may introduce distractions or, worse, actually distort the data, making it harder to decode the data accurately.

Effectiveness of a graph can be enhanced by removing unnecessary artifacts from the graph. Avoid usage of gradient background and images. Inclusion of embellishments like drop shadows for data markers can increase the visual appeal of the graph but can reduce the effectiveness of the graph.

1.1.6 Short-Term Memory

Generally, people find it difficult to absorb and retain a large number of data values at one time. Short-term memory is limited. Arranging the data in smaller chunks can aid in the processing of information. Research in this field shows:

- An average person can easily remember 3–5 chunks of information.
- An average person can mentally calculate with 2-digit numbers.
- Excessive eye movement reduces efficiency of decoding a graph.

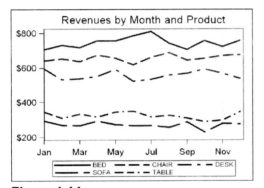

Figure 1.14   **Figure 1.15**

Both of the graphs above display revenues by product over time. The data are grouped by product, each series representing one product. Figure 1.14 uses a traditional legend at the bottom of the graph to identify each product.

To compare revenues for desks and chairs, you have to move your eyes down to the legend and then back up to the plot. Figure 1.15 uses direct labeling for each series. This eliminates eye movement and thus facilitates easier comparisons of the data.

1.1.7 Summary

You can use the above-mentioned guidelines to create graphs that convey information with maximum effectiveness and minimum distractions. In summary, we suggest that you:

- Use linear distances from a common baseline to represent magnitude.
- Use marker shapes or line patterns for grouping.
- Simplify the appearance of the plot and reduce unnecessary ink.
- Increase proximity of data for better comparisons.
- Reduce eye movement needed to decode the data.

The SG Procedures provide the features you need to implement the above guidelines.

1.2 Automatic Graphs from SAS Procedures

Now let us see how you can get effective graphs using SAS. Starting with SAS 9.2, many SAS analytical procedures use the Output Delivery System (ODS) Graphics system to automatically produce graphs. When the ODS Graphics system is enabled, such graphs are produced along with the data tables in the right order, and included in the output file such as a PDF document.

To create the graphs from SAS 9.2 analytical procedures, you only need to switch on the ODS Graphics system before running the procedure. This is done by including the following statements in your program. No additional graph code is required.

```
ods graphics on / < options >;
   < procedure statements; >
ods graphics off;
```

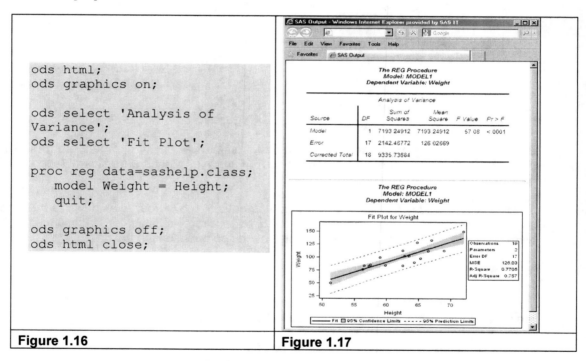

```
ods html;
ods graphics on;

ods select 'Analysis of
Variance';
ods select 'Fit Plot';

proc reg data=sashelp.class;
   model Weight = Height;
   quit;

ods graphics off;
ods html close;
```

Figure 1.16   **Figure 1.17**

Figure 1.16 shows the usage of ODS Graphics with an example of the REG procedure. In this example, we have specifically requested the output of the Analysis of Variance table and the Fit Plot by using the ODS SELECT statements. The resulting HTML output is shown in Figure 1.17. The Analysis of Variance table and the Fit Plot are produced in the right sequence in the output HTML file.

It is worth repeating that the Fit Plot is produced automatically, without any graph coding required on the part of the user. It is useful to note that the procedures have to run additional processing steps to create these graphs. Some procedures may create a large set of graphs, which is something to consider when using this option.

With SAS 9.3, running in DMS mode, the default open destination is HTML, and ODS Graphics is on by default. For line mode, the default open destination is LISTING, and ODS Graphics is off by default. This is a change from SAS 9.2, where the default open destination is LISTING, and ODS Graphics is off by default.

1.3 Graph Template Language

The graphs mentioned in section 1.2 are created automatically by the procedure and do not require any additional programming by user. This is done by using predefined graphics templates that use the Graph Template Language (GTL). All templates needed to create these graphs have been supplied with the software. The graph shown in Figure 1.17 is created by the REG procedure using one of the predefined templates shipped with SAS.

GTL can also be used directly by you, the SAS user, to create your own custom graph template. Then, you can use the SGRENDER procedure to associate this template with the appropriate data to produce the resultant graph.

A detailed description of GTL and the SGRENDER procedure is beyond the scope of this book. However, it should be noted that all graphs created by ODS Graphics system are done using the GTL syntax at some level. This is also true of the graphs produced by the Statistical Graphics (SG) procedures, which is the topic of this book.

1.4 Statistical Graphics Procedures

The SG procedures are a set of procedures that work within the ODS Graphics system. These procedures are designed to provide the familiar procedure syntax for creation of graphs that are most commonly used in various industries such as health and life sciences, finance, banking, quality control, and more.

The graphs are created using GTL behind the scene. So, these graphs have the same look and feel as the automatic graphs created from the SAS analytical procedures. These graphs are useful for visualization of the raw data or for custom graphs of analysis results.

1. **Pre-analysis Data Exploration**
    When you receive the data from a survey or a study, often you may want to get multiple graphical views of the raw data before the analysis phase. Viewing the data using simple scatter plots, histograms, and scatter plot matrices can provide the analyst valuable insights into the data which can help in the analysis phase of the project.

2. **Analysis of Data**

    This phase of the project involves the analysis of the data using analytical procedures and/or your own custom data steps. Automatic graphs can be obtained from individual procedures as mentioned in section 1.2. In this phase you may also need to create specialized graphs that are not currently supported by the procedure itself.

3. **Post-analysis Data Presentation**

    After the analysis phase of the project, you may need to present the results of the analysis in a form that is easily consumed by your audience. This goal is best achieved by presenting the results in the form of graphs that include the original data along with the analytical results. If the analysis requires multiple-procedure steps or custom data steps, it may be necessary to create custom graphs from the results.

For all steps in the process above, you need the ability to create graphs from the raw data or from the results of your custom analysis. You may also want to use the graphs that are automatically created for you by individual procedures in the report for the project. In this case, the SG procedures are the ideal tools for this job for the following reasons:

- Graphs created by the SG procedures are identical in look and feel to the automatic graphs created by the analytical procedures. Mixing and matching the output from the SG procedures and the analytical procedures is seamless.

- SG procedure steps can be run along with the analytical procedures and data steps to produce a sequential output in the open ODS destination.

- SG procedures provide a simple and concise syntax to create many types of graphs, classification panels, and scatter plot matrices.

- With SAS 9.3, SG procedures also provide the ability to annotate the graph with Annotate-like functionality using a data set. Additionally, attribute maps can be used to control the usage of visual attributes like color or marker symbols in the graph. These topics will be discussed in detail in Chapter 9.

1.5 Organization of This Book

The approach taken for this book is to present the features of the SG procedures via examples. A textbook on this topic would normally take you through all the features of the procedures, and you would have to know many aspects of the procedures to be able to generate a useful graph.

Instead of listing all the options and features of the procedures, we take the reverse approach. If you have an idea of the graph you want to make, you can just flip through this book and find the graph closest to what you need. Then, right alongside, you will find the code necessary to create the graph. From there, you can build on the graph by borrowing from other examples in the book.

SG procedures utilize a building-block approach to creating a graph. If you see two graphs that each individually include elements that you want in one graph, it is highly likely that you can combine the statements in one procedure step and get the combined graph you need.

For example, you can combine a scatter plot, a series plot, and various regression plot statements from different examples into one procedure step. Some common combinations are as follows:
- Scatter, Series, Step, Band, Regression, Ellipse, VBarParm, and HBarParm
- Histogram and Density
- Bar Chart, Line Chart, and Dot Plot

This book is organized as follows:

- In Chapter 2, we start with a general description of each procedure. This will show you the structure of the syntax and the main features with a few examples. From there on, we focus on examples, starting with single-cell graphs, and then moving on to more complex cases.

- In Chapter 3, we review graphs that are commonly used in various domains. This section covers the different graph types you can create. We defer the detailed discussion of the plot options to subsequent chapters.

- In Chapters 4–7, we cover the main groups of single-cell graphs using the SGPLOT procedure. Plot statements used in these graphs can be combined within the groups to create the graph you need. Various supported options are used to demonstrate the features.

- In Chapter 8, we cover common customizations for axes, legends, and insets.

- In Chapter 9, we cover the topics of annotation and attribute maps. Annotations allow you to add custom graphical elements to a graph that may or may not be data driven. Attribute maps provide you the ability to tie the plot attributes like color, symbols, and line patterns to explicit data values. These powerful features for detailed customization of graphs are included with SAS 9.3.

- In Chapter 10, we cover classification panels using the SGPANEL procedure. This topic leverages all you have learned about single-cell graphs to produce graphs that are classified by multiple class variables.

- In Chapter 11, we cover comparative scatter plots and scatter plot matrices using the SGSCATTER procedure.

- In Chapter 12, we cover graphs commonly used in the health and life sciences industry.

- In Chapter 13, we cover some special business graphs. Here you will find detailed examples that combine features from previous chapters to create the graph.

- In Chapter 14, we cover the topic of styles. Here you will see the inner workings of styles and the association between style elements and graph features. We will cover the basics of creating your own custom style for graphs.
- In Chapter 15, we cover the options on the ODS DESTINATION statement that have a direct bearing on the rendering of the graphs. We also review the options you can set on the ODS GRAPHICS statement to control aspects of graph rendering.
- In Chapter 16, we cover how to create graphs appropriate for different use cases. Often, graphs are created for inclusion in a full slide of a Microsoft PowerPoint presentation, or in one 3-1/4" column of a Microsoft Word document for a printed journal. We will provide some tips on how to create graphs that are suitable for such use cases.

1.6 Data Sets and Custom Styles

Most of the examples in this book use the pre-defined SAS data sets available in the SASHELP library. These include CARS, HEART, and a few others. Often, to reduce the number of classifiers, so that a graph will fit in the restricted space, modified data sets are used that contain a subset of the data from these data sets. In other cases, custom data sets are needed that are suitable for the example graph.

Custom styles are sometimes used to render some of the graphs in this book. Primarily, these are necessary to reduce the font sizes to help fit the graphs into the small space available. The results you see may vary based on the active style for an ODS destination.

1.7 Color and Gray-Scale Graphs

The graphs created by the SG procedures, and by the ODS Graphics system in general, use the active style of the open destination. Often, these styles are optimized for full color output. This works well when the graph is also consumed in a color medium.

However, when color graphs are printed in gray scale, there is a significant loss of fidelity in the representation of distinct categories in the graph. For example, a graph with two series plots, one for Drug A and one for Drug B can be well represented in color with use of two distinct colors, say red and blue. These colors are often designed to have equal weight to avoid unintentional bias.

When such a graph is printed in gray scale, these two series plots may look very similar unless they have other distinguishing features such as line patterns and marker shapes to facilitate discrimination between groups. Bar charts can benefit from use of fill patterns to facilitate such discrimination.

This book is printed in gray scale, so it is important to create the graphs that will print well in a gray-scale format. To ensure this, it is best to create the original graph in the gray-scale format that maximizes the discriminability of the different categories and groups. All of the graphs included in this book are created using gray-scale styles such as Journal, Journal2 or Journal3, or styles derived from these styles.

When you run a program from this book, or one of your own, the graph will be rendered using the active style of the open destination. For SAS 9.2, this is the LISTING destination. For SAS 9.3 in DMS session, this is the HTML destination. Since both of these destinations use a default color style, you will get a graph rendered in full color. To get a gray-scale graph, use one of the styles mentioned above.

The WIDTH or HEIGHT options on the ODS GRAPHICS statement have been used to render the graphs for this book. However, these options are not shown in the sample code. When you run the same code without these options, the graphs will render in the default size.

1.8 Effective Graphics and the Use of Decorative Skins

As alluded to earlier, the SG procedures specifically, and ODS Graphics in general, are designed with the principles of effective graphics in mind. By default, the procedures always strive to create a graph that delivers the information with maximum clarity and minimum clutter or distraction.

Though initially designed with the statistical user in mind, these procedures are finding increasing usage in non-statistical domains. In these use cases, there is often a desire for a flashier graph, even at the expense of effectiveness.

In the SG procedures, the bar chart statements support an option to apply a decorative skin to a bar. This does not change the shape of the bar but provides a "flashier" rendering. This option can be used at the discretion of the user. Some examples in this book use this option to demonstrate this feature. The intention is primarily to expose these available features.

1.9 SAS 9.2 and SAS 9.3 Features

This book assumes the user has the SAS 9.2 (TS2M0) or higher release. The examples in this book include the usage of SAS 9.2 (TS2M0) and SAS 9.3 features. Examples that use the features added in SAS 9.3 are marked with the 9.3 icon, and the specific statement or option is displayed in **bold italics**.

In some cases you can remove the new option and still run the code using SAS 9.2.

Note: When running SAS 9.3 in DMS mode, the default open destination is HTML. For non-DMS mode, the default open destination is LISTING. The ODS GRAPHICS feature is automatically enabled for the execution of the SG procedures.

16

Chapter 2

Statistical Graphics Procedures

2.1 Key Concepts   19
2.2 SGPLOT Procedure   22
2.3 SGPANEL Procedure   24
2.4 Combining Statements   29
2.5 SGSCATTER Procedure   30
2.6 Styles and Their Usage   32
2.7 Template-Based Graphics vs. Device-Based Graphics   33

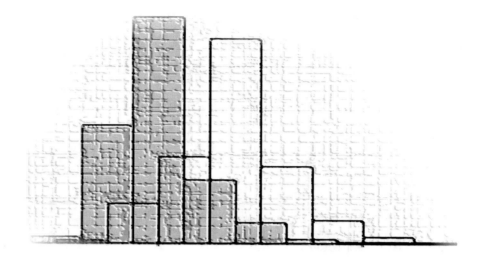

*"Simple things should be simple.
Complex things should be possible."
~ Alan Kay*

Chapter 2: Statistical Graphics Procedures

In this chapter we introduce the key concepts for the SG procedures and the general syntax.

The SG procedures provide a direct procedure interface into the ODS Graphics system. These procedures create graphs with very little code. The SGPLOT and SGPANEL procedure syntax allows you to build up complex graphs by use of plot statements and other features. The SGSCATTER procedures syntax is designed to give single-statement access to a variety of scatter plot panels and matrices.

2.1 Key Concepts

The graphs created by these procedures can be grouped into three broad categories: single-cell graph, multi-cell classification panels, and multi-cell comparative scatter plots. These categories are discussed in the following sections.

2.1.1 Single-Cell Graph (SGPLOT Procedure)

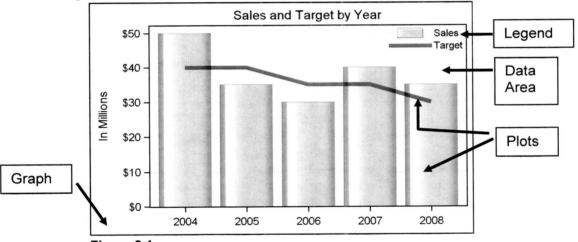

Figure 2.1

Figure 2.1 shows a typical single-cell graph built using the following components:
- one or more titles at the top of the graph.
- one or more footnotes at the bottom of the graph
- one region in the middle displaying the data
- one or more plots in the data area
- one or more legends or insets inside the data area or outside

Note: All statements are referred to as "plot", regardless of whether it is a series plot, scatter plot, bar chart, histogram, and so on.

Graph: Refers to the individual output created by the procedure. In most of the common use cases, each execution of the procedure creates one graph output file. Often these procedures produce multiple output files (for BY variable usage, or paging of large panels), each of which is referred to as a "Graph".

Cell: Each graph can have one or more data areas where the data is plotted as shown in Figure 2.2. Each of these is referred to as a "Cell". A cell may or may not include axes.

Plot statements: Each plot statement is responsible for drawing only its own data representation. It is told by the container where to draw itself and how to scale its data.

Axes: The X and Y axes are shared by all the plots in the graph. The data range for each axis is determined by the graph and is based on all the plots placed in it. Each graph can have a second set of axes, called X2 (at the top) and Y2 (on the right). Each plot can specify which axes it wants to use.

Legends: A graph can have one or more legends, and each can be placed in any part of the graph. Each legend can specify the information to be displayed in it.

Single-cell graphs are created using the SGPLOT procedure.

2.1.2 Multi-Cell Classification Panels (SGPANEL Procedure)

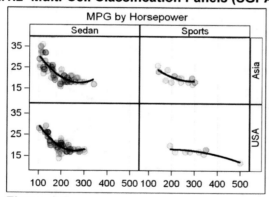
Figure 2.2

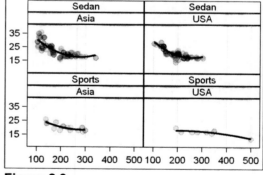
Figure 2.3

Classification panels are very useful to visualize the distribution of data classified by one or more class variables in one display. Figures 2.2 and 2.3 both display the association between MPG and Horsepower for vehicles by country of origin and type of car. From both graphs, one can easily glean some information:

- For both Asian and US cars, "Sedan" outnumbers "Sports" cars.
- Figure 2.2 shows that US sports cars provide higher horsepower choices.
- Figure 2.3 shows that Asian cars offer higher MPG choices.

Figure 2.2 displays a classification panel with a Lattice layout, which supports two class variables, one for row and one for column. Each row and column has a header that displays the value of the classification variables.

Figure 2.3 displays a classification panel with a Panel layout, which supports multiple class variables. Each cell in the panel has multiple headers, one for each class variable.

Classification panels are created using the SGPANEL procedure.

2.1.3 Multi-Cell Comparative Scatter Plots (SGCATTER Procedure)

While Classification Panels provide a convenient way to compare the same data across classifiers, it is often desirable to make side-by-side comparisons of different measures.

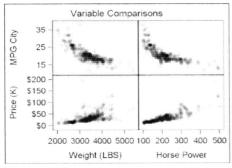

Figure 2.4   **Figure 2.5**

Figure 2.4 displays a comparative graph for Mileage and Price by Weight and Horse Power. This graph has common axes for comparison of the values. Figure 2.5 shows a Scatter Plot Matrix for three variables, MPG, Horsepower, and MSRP. Such a matrix can provide preliminary visual indication of direct or inverse associations between the variables.

Comparative and Matrix graphs are created using the SGSCATTER procedure.

2.1.4 Automatic Features

The SG Procedures examine the procedure syntax and apply built-in heuristics to enhance the graph automatically. These actions include the following:
- Add a legend when appropriate for multiple overlays and classifiers.
- Create custom legend labels as needed for certain plot types.
- Assign different visual attributes to overlaid plot statements.
- Paginate large classification panels automatically.

If the results of the built-in heuristics are not desirable, they can be turned off.

2.2 SGPLOT Procedure

SGPLOT is a versatile procedure, providing the ability to create a wide variety of single-cell graphs by combining the supported statements in creative ways. The SGPLOT procedure supports over twenty different plot statements, along with statements for customization of legends, axes, and insets.

The syntax for the SGPLOT procedure is as follows:

```
1. PROC SGPLOT < DATA= data-set > < options >;
2. plot-statement(s) required-parameters < / options >;
           ⎧  < refline-statement(s) >;
3.         ⎨  < inset-statement(s) >;
           ⎪  < axis-statement(s) >;
           ⎩  < keylegend-statement(s) >;
   RUN;
```

1. The procedure statement supports multiple options. Use of these options will be demonstrated in the examples shown in later chapters.

2. One or more plot statements are used to represent the data. Each plot statement has its own set of required data roles and options. These are described in detail in later chapters. Many plot statements are supported and can be grouped as shown below. Plot statements from compatible groups can be combined in one procedure step:

   a. **Basic Plots:** These include scatter, series, etc.
   b. **Fit and Confidence Plots:** These include regression and Loess plots.
   c. **Distribution plots:** These include histograms and box plots.
   d. **Categorization Plots:** These include bar charts and dot plots.

3. Supporting statements can be used to customize the graph:

   a. Reference Lines
   b. Insets
   c. Axes
   d. Legends

Figure 2.6 shows a typical use case of the SGPLOT procedure. In this example, we have created a distribution plot for the variable "Horsepower". Three separate plot statements are used, one for each data element of the graph. These statements are rendered in the order in which they are specified as illustrated in Figure 2.7. Figure 2.9 shows the actual rendered graph.

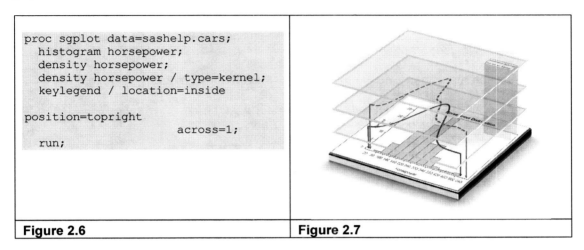

| **Figure 2.6** | **Figure 2.7** |

In Figure 2.6, we start with the HISTOGRAM statement, and then place two DENSITY plots on it. The KEYLEGEND statement is used to customize the position of the legend.

- The HISTOGRAM statement is placed first in the area bounded by the axes.
- The first DENSITY curve ("Normal" by default) is placed on top of the histogram.
- The second DENSITY curve (Type=Kernel) is placed on top of the Normal curve.
- The legend is actually built by default by the procedure and placed on top. The KEYLEGEND statement is used to customize its location.

Figure 2.8 and Figure 2.9 display the program and the resulting graph. The XAXIS statement is added to suppress the x-axis label, as it is unnecessary given the title of the graph.

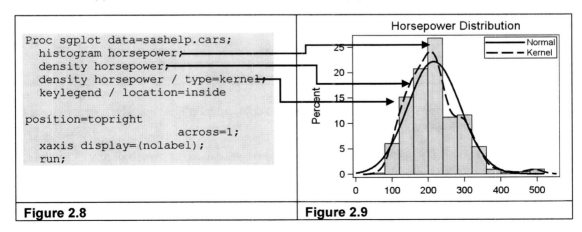

| **Figure 2.8** | **Figure 2.9** |

As mentioned above, the supported plot statements are grouped in four categories. Plots within each category can be combined in a procedure step. Plots from the "Basic Plots" group and "Fit and Confidence Plots" group can also be combined.

INSET, REFLINE, and KEYLEGEND statements can be used as necessary to add these elements to your graph.

Axes statements can be used to customize the appearance of any of the four axes:

- XAXIS is along the bottom of the graph.
- YAXIS is along the left side of the graph.
- X2AXIS is along the top of the graph.
- Y2AXIS is along the right side of the graph.

2.3 SGPANEL Procedure

The SGPANEL procedure extends the features of the SGPLOT procedure to a classification panel, using one or more classification variables. The examples shown earlier in Figure 1.12 and Figure 1.13 are classification panels created using the SGPANEL procedure.

The syntax for the SGPANEL procedure is as follows:

```
1.  PROC SGPANEL < DATA= data-set > < options >;
2.    PANELBY classvar1 < classvar2 … <classvarN > < / options >;
3.    < plot-statement(s) >;
4.    < refline-statement(s) >;
      < inset-statement(s) >;
      < axis-statement(s) >;
      < keylegend-statement(s) >;
    RUN;
```

1. The procedure statement supports multiple options that will be demonstrated in the examples shown in later chapters.

2. The PANELBY statement is required and must be placed before of any of the plot, refline, inset, axis, or legend statements. This statement is used to set the layout type and other options that control the overall paneling of the cells.

3. One or more plot statements are used to represent data. Each plot statement has its own set of data roles and options described in detail in later chapters. This procedure supports most of the same plot statements as the SGPLOT procedure.

4. Supporting statements can be used to customize the graph:
   a. Reference Lines
   b. Axes
   c. Legends

Figure 2.10 shows program statements for the creation of a classification panel displaying the distribution of mileage by origin and type. The resulting graph is shown in Figure 2.11.

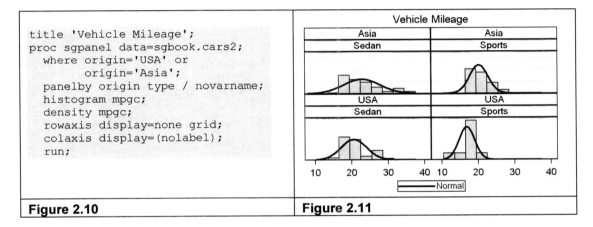

Figure 2.10   **Figure 2.11**

The program code in Figure 2.10 has the following noteworthy items:

- The PANELBY statement is required, and must come before any plot, axes, reference line, or legend statements. This sets the classification variables.
- The two plot statements, HISTOGRAM and DENSITY, define the "Prototype" that is used to populate each cell of the panel.
- The ROWAXIS and COLAXIS statements are used to customize the row and column axes.

The PANELBY statement must provide one or more class variables. The procedure automatically subdivides the graph region into multiple cells based on all crossings of the unique values of the class variables. In Figure 2.10, we have specified two class variables "Origin" and "Type", each having two unique values. Hence, we get a graph with four cells, as shown in Figure 2.11. Each cell has two headers.

Each cell in the graph is populated with the same set of plot statements as specified in the procedure syntax. In this case, each cell gets a HISTOGRAM and a DENSITY plot. The data for each cell is a subset based on the class value(s) for the cell. In Figure 2.11, the cell titled "Asia" displays the subset of the data where Origin='Asia'.

The non-plot statements such as the ROWAXIS and COLAXIS are used to customize the common axes. A legend is automatically created based on an inspection of the syntax.

All cells of the graphs created by this procedure have common external axes:

- o The x-axes for all columns are uniform.
- o The y-axes for all rows are uniform.
- o Uniform option may be set within rows or columns.

The SGPANEL procedure supports a number of layout types, which will be described in the following sections.

2.3.1 Layout PANEL

This is the default layout type. This layout supports one or more (N) class variables. A cell is created for each crossing of the unique values of all the class variables. Figure 2.12 shows a graph with two class variables, each with 2 levels, resulting in a graph with 4 cells.

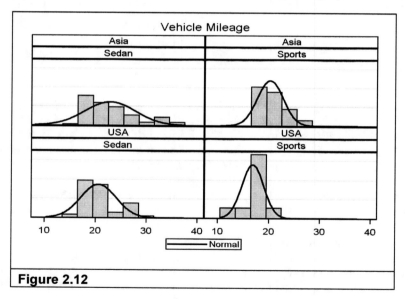

Figure 2.12

A graph with a panel layout has the following features:

- Each cell has N cell headers, one for each class variable, displaying the value for each cell.
- By default, only the cells that have data are displayed. Cells without any data are dropped from the graph. An option can be used to display all cells.
- The procedure automatically decides the number of rows and columns for the grid. When a graph has many cells, the procedure will automatically break up the graph into multiple "pages", to prevent the cells from getting too small.
- Common external row and column axes are used.
- Options are available to allow the user to control the "paging" of the graph.

2.3.2 Layout LATTICE

This layout requires two class variables. The first class variable is treated as the Column variable, and the second as the Row variable. See Figure 2.13.

- Each unique value of the Column variable creates a column in the grid.
- Each unique value of the Row variable creates a row in the grid.
- Each column gets a column header, by default at the top.
- Each row gets a row header, by default on the right.
- Common external row and column axes are used.
- Every crossing of the unique values of the column and row variables is displayed as a cell, regardless of whether it has data or not.

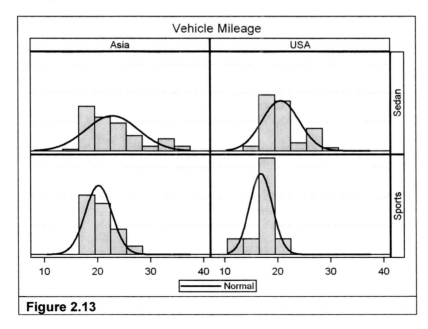

Figure 2.13

2.3.3 Layout COLUMNLATTICE

This is a special case of the LATTICE layout, where only one variable is provided in the list of class variables. This layout produces a panel of columns in one row. Each cell has a column header as shown in Figure 2.14.

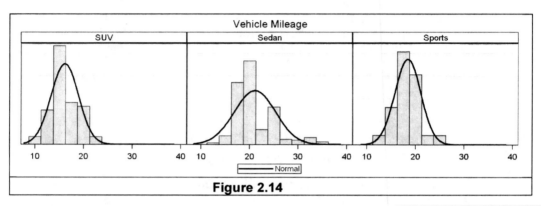

Figure 2.14

2.3.4 Layout ROWLATTICE

This is a special case of the LATTICE layout, in which only one variable is provided in the list of class variables. This layout produces a panel of rows stacked in one column. Each cell has a row header as shown is Figure 2.15.

2.3.5 Automatic Paging

Classification panels provide a convenient way to visualize data by multiple class variables. This works very well when the number of cells in a panel is small, say, less than ten. It is possible to display a grid of 3 x 3 cells in a graph on a page.

However, as the number of cells grows larger, the individual cell size shrinks. At some point, each cell becomes too small to reasonably display the data contents. The SGPANEL procedure automatically manages the number of cells in a graph to keep reasonable cell sizes. If needed, the graph is "paged" into multiple images, each showing a subset of the full panel.

Options are available for user control of such paging, as will be shown in some examples later in the book.

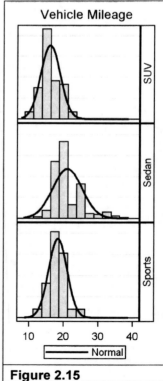

Figure 2.15

2.4 Combining Statements

Plot statements from the Basic Plots and Fit and Categorization Plots groups can be mixed and matched together freely. For other plot statements, only specific combinations are allowed. The table in Figure 2.16 below shows the allowable combinations.

* Ellipse is not supported by PROC SGPANEL.

	SCATTER	SERIES	STEP	BAND	NEEDLE	VECTOR	VBARPAR	HBARPAR	BUBBLE	HIGHLOW	REFLINE	LINEPAR	REG	LOESS	PBSPLINE	ELLIPSE *	HISTOGR	DENSITY	VBOX	HBOX	VBAR	VLINE	HBAR	HLINE	DOT
Basic Plots																									
SCATTER	x	x	x	x	x	x	x	x	x	x	x	x	x	x	x	x									
SERIES	x	x	x	x	x	x	x	x	x	x	x	x	x	x	x	x									
STEP	x	x	x	x	x	x	x	x	x	x	x	x	x	x	x	x									
BAND	x	x	x	x	x	x	x	x	x	x	x	x	x	x	x	x									
NEEDLE	x	x	x	x	x	x	x	x	x	x	x	x	x	x	x	x									
VECTOR	x	x	x	x	x	x	x	x	x	x	x	x	x	x	x	x									
VBARPARM	x	x	x	x	x	x	x		x	x	x	x													
HBARPARM	x	x	x	x	x	x		x	x	x	x	x													
BUBBLE	x	x	x	x	x	x	x	x	x	x	x	x	x	x	x	x									
HIGHLOW	x	x	x	x	x	x	x	x	x	x	x	x	x	x	x	x									
REFLINE	x	x	x	x	x	x	x	x	x	x	x	x	x	x	x	x	x	x	x	x	x	x	x	x	x
LINEPARM	x	x	x	x	x	x	x	x	x	x	x	x	x	x	x	x									
LINEPARM	x	x	x	x	x	x	x	x	x	x	x	x	x	x	x	x									
Fit and Confidence Plots																									
REG	x	x	x	x	x	x			x	x	x	x	x	x	x	x									
LOESS	x	x	x	x	x	x			x	x	x	x	x	x	x	x									
PBSPLINE	x	x	x	x	x	x			x	x	x	x	x	x	x	x									
ELLIPSE *	x	x	x	x	x	x			x	x	x	x	x	x	x	x									
Distribution Plots																									
HISTOGRAM											x						x	x							
DENSITY											x						x	x							
VBOX											x								x						
HBOX											x									x					
Categorization Plots																									
VBAR											x										x	x			
VLINE											x										x	x			
HBAR											x												x	x	x
HLINE											x												x	x	x
DOT											x												x	x	x

Figure 2.16

2.5 SGSCATTER Procedure

The SGSCATTER procedure is optimized for the creation of comparative scatter plots and scatter plot matrices. PROC SGSCATTER does not use a layered architecture like the SGPLOT and SGPANEL procedures. Instead, PROC SGSCATTER supports three distinct plot statements as shown below. The examples shown earlier in Figure 2.4 and Figure 2.5 are created using the SGSCATTER procedure.

The syntax for the SGSCATTER procedure is as follows:

```
1   PROC SGSCATTER < DATA= data-set > < options >;
2.  PLOT  plot-requests < / options >;
3.  COMPARE X=(variable(s)) Y=(variable(s)) < / options >;
4.  MATRIX variable(s) < / options >;
5.  RUN;
```

1. The procedure statement supports multiple options that will be demonstrated in the examples shown in later chapters.

2. The PLOT statement creates a set of cells arranged in a uniform grid in the graph based on the plot request. Each cell includes an independent scatter plot with optional fit and ellipses. Axes can be made uniform across all plots, if needed. Figures 2.17 and 2.18 show an example of the PLOT statement. The two Y axes are not uniform.

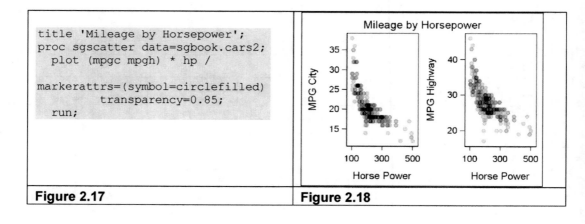

```
title 'Mileage by Horsepower';
proc sgscatter data=sgbook.cars2;
  plot (mpgc mpgh) * hp /
markerattrs=(symbol=circlefilled)
         transparency=0.85;
  run;
```

Figure 2.17   **Figure 2.18**

3. The COMPARE statement creates a row and column grid of cells in the graph based on the list of X and Y variables. Each cell includes a scatter plot with optional fit and ellipses. Each row of the grid has a common external y-axis. Each column of the grid has a common external x-axis. Options can be used to customize the plots. Figures 2.19 and 2.20 show an example of the COMPARE statement.

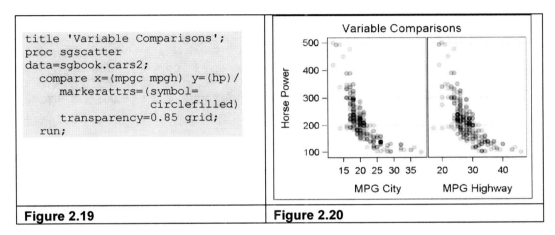

```
title 'Variable Comparisons';
proc sgscatter
data=sgbook.cars2;
  compare x=(mpgc mpgh) y=(hp) /
    markerattrs=(symbol=
                 circlefilled)
    transparency=0.85 grid;
run;
```

Figure 2.19   **Figure 2.20**

4. The MATRIX statement creates a row and column grid of cells based on the list of variables specified. Each cell includes a scatter plot with optional fit and ellipses. Each row and column of the grid has external axes. Figures 2.21 and 2.22 show an example of the MATRIX statement.

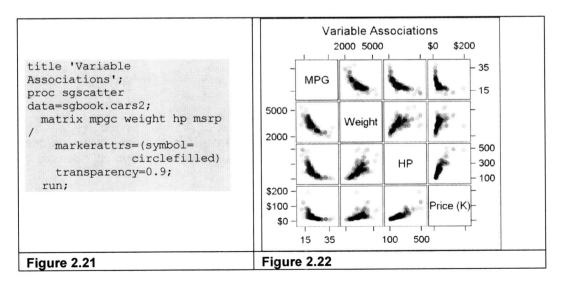

```
title 'Variable
Associations';
proc sgscatter
data=sgbook.cars2;
  matrix mpgc weight hp msrp
/
    markerattrs=(symbol=
                 circlefilled)
    transparency=0.9;
  run;
```

Figure 2.21   **Figure 2.22**

2.6 Styles and Their Usage

Graphs created by the SG procedures use pre-defined visual attributes for various elements that make up a graph. This information is defined in ODS Styles that have been carefully designed with aesthetics and effectiveness in mind. This ensures these graphs have an aesthetically pleasing appearance by default. This also ensures that various elements of the graph are balanced and distinct from each other. This is achieved through the usage of the ODS Styles.

A style is a collection of named elements. Each element is a bundle of named visual attributes such as color, font, marker symbol, and so on. Any output to an ODS destination derives the necessary visual information from the active style. For example, output tables derive visual attributes such as fonts and background colors for the headers, size, and color of titles, etc. from the style.

Graphs also derive the visual attributes for plot colors, marker symbols, line thickness, axis label fonts, etc. from specific named elements of the style. The association between the element of the graph and the style element is well defined and described in detail in the ODS product documentation. Some relevant information is included in Chapter 14.

You can control the visual appearance of the graphs in different ways:

1. Use a pre-defined SAS style.
2. Use a custom style.
3. Use appearance options.

1. Use a pre-defined SAS style: Every ODS output destination has a default style. All graphs written to this destination use this style by default. You can change the active style for an ODS output destination by setting the STYLE= option for the destination. All graphs written to that destination will then use that style.

 ods *destination* style=*style-name*;

2. Use a custom style: If you like one of the pre-defined SAS styles, but would prefer to change a few of the appearance settings, you can derive a new style from one of the SAS styles by using the TEMPLATE procedure. For more information on this topic, see the ODS product documentation for PROC TEMPLATE.

3. Use appearance options: A style set on the ODS destination affects all features of the all graph rendered to the destination. You can customize the appearance of specific features of a graph by setting specific appearance options in the procedure syntax. This overrides the settings derived from the style only for that one use case.

Usage of these options will become apparent through many examples in this book. More information on this topic is covered in Chapter 14.

2.7 Template-Based Graphics vs. Device-Based Graphics

The SG procedures work within the ODS Graphics system and use GTL to do the work of rendering the graph output. For this reason, we will refer to these procedures as "template-based graphics". The other SAS/GRAPH® procedures, such as GPLOT and GCHART, will be referred to as "device-based graphics".

Often, users ask what is different about these procedures from the other SAS/GRAPH® procedures. While we cannot cover every difference here, the key differences in philosophy, process, and technology are listed below:

1. SG procedures create output directly in industry standard formats, such as PNG, JPG, PDF, PS, and so on. SG procedures do not create GSEG output.

2. SG procedures do not honor traditional GRAPH global statements such as GOPTIONS, PATTERN, SYMBOL, AXIS or LEGEND. However, equivalent or similar functionality is provided through other procedures, ODS Graphics, or ODS destination options.

   a. The GOPTIONS statement is not supported (except for using the RESET option to reset titles and footnotes). Global options that control the output are specified on the ODS GRAPHICS statement.

   b. TITLE and FOOTNOTE statements are supported, with the exception of a few options.

   c. PATTERN and SYMBOL statements are not supported. However, plot types and attributes are specified directly on the plot statements, which make the application of those attributes clearer.

   d. AXIS statements are not supported. However, PROC SGPLOT supports four axis statements (XAXIS, YAXIS, X2AXIS, and Y2AXIS) to control the axes around the cell. PROC SGPANEL supports the ROWAXIS and COLAXIS statements to control its axes. PROC SGSCATTER does not have axis statements, but there are techniques that will be described later for controlling some aspects of the axes.

   e. LEGEND statements are not supported. PROC SGPLOT and SGPANEL support a KEYLEGEND statement to control the contents of the legend, its layout, and position. PROC SGSCATTER has a LEGEND option to give some legend control.

Chapter 3

Common Graphs

3.1   Introduction   37
3.2   Single-Cell Graphs   38
3.3   Classification Panels   55
3.4   Comparative and Matrix Graphs   57

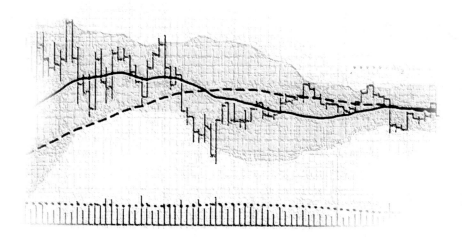

"There are 10 types of people in this world:
those who understand binary and those who don't."
~ Unknown

Chapter 3: Common Graphs

In this chapter, we review some graphs commonly used in many domains, from the basic scatter plot to complex comparative graphs and classification panels. We will defer the detailed discussion of various plot options to subsequent chapters.

3.1 Introduction

All the graphs have a common structure:

- one or more titles at the top and one or more footnotes at the bottom
- various graphical representations of the data in the middle
- one or more legends, inside or outside the data area
- insets or statistics table inside or outside the data area

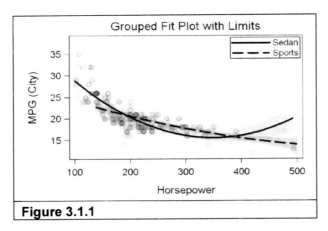

Figure 3.1.1

Single-Cell Graphs: In section 3.2, we will use the SGPLOT procedure to create single-cell graphs. We will put together the needed syntax to create graphs like the one shown above in Figure 3.1.1.

Classification Panels: In section 3.3, we will use the SGPANEL procedure to create classification panels. Once you know how to mix and match statements for PROC SGPLOT, all you need is a PANELBY statement to set up the classification variables.

Comparative and Matrix Graphs: In section 3.4, we will use the SGSCATTER procedure to create comparative and matrix graphs.

In the rest of this chapter, we provide a broad coverage of the different types of graphs you can create using the SG Procedures. The goal is to expose to you the different types of graphs, without going deep into the features of each plot. We will cover that in subsequent chapters.

3.2 Single-Cell Graphs

Figure 3.2.1: Basic Scatter Plot

This is a basic scatter plot of mileage by horsepower for sedans and sports cars.

Default option settings are used, and the appearance settings are derived from the active style.

Axes are thinned, and "optimal" tick values are displayed.

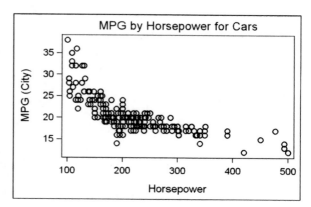

```
title 'MPG by Horsepower for Cars';
proc sgplot data=sashelp.cars;
  where type='Sedan' or type='Sports';
  scatter x=horsepower y=mpg_city;
  run;
```

Figure 3.2.2: Grouped Scatter Plot with Grid Lines

This is a grouped scatter plot of mileage by horsepower by car type. The visual attributes for each marker are based on the unique value of the group variable.

The legend is positioned in the upper right corner of the data area using the KEYLEGEND statement.

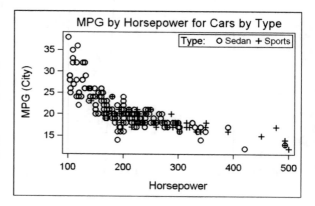

```
title 'MPG by Horsepower for Cars by Type';
proc sgplot data=sashelp.cars;
  where type='Sedan' or type='Sports';
  scatter x=horsepower y=mpg_city/ group=type;
  keylegend / title='Type: ' location=inside position=topright;
  xaxis grid;
  yaxis grid;
  run;
```

Figure 3.2.3: Fit Plot

This is a fit plot of mileage by horsepower. The REG statement displays both the original data and the fit line. The large number of observations can obscure the fit line and clutter up the graph.

The thickness of the fit line is set using the LINEATTRS option.

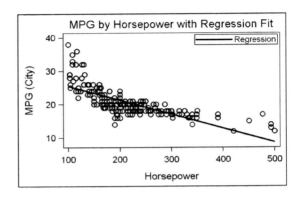

```
title 'MPG by Horsepower with Regression Fit';
proc sgplot data=sashelp.cars;
  where type='Sedan' or type='Sports';
  reg x=horsepower y=mpg_city / lineattrs=(thickness=2);
  keylegend / location=inside position=topright;
  xaxis grid;
  yaxis grid;
  run;
```

Figure 3.2.4: Fit Plot with Transparent Markers

In this fit plot, the clutter is reduced by suppressing the markers on the REG statement and using a SCATTER statement to display the markers with transparency.

Using transparency can provide a better feel for the distribution of the data.

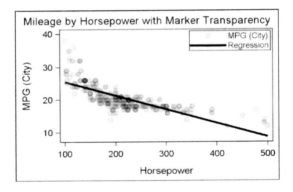

```
title 'Mileage by Horsepower with Marker Transparency';
proc sgplot data=sashelp.cars;
  where type='Sedan' or type='Sports';
  scatter x=horsepower y=mpg_city / transparency=0.9
          markerattrs=(symbol=circlefilled);
  reg x=horsepower y=mpg_city / nomarkers lineattrs=(thickness=3);
  keylegend / location=inside position=topright across=1;
  xaxis grid;
  yaxis grid;
  run;
```

Figure 3.2.5: Fit Plot with Confidence Limits

This fit plot uses a regression plot with 2nd order fit using the DEGREE option on the REG statement. Confidence limits are requested using the CLI and CLM.

The default legend is positioned in the top right corner of the graph.

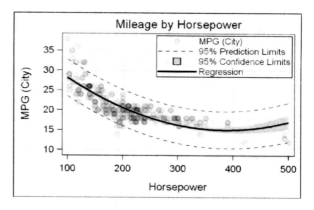

```
title 'Mileage by Horsepower';
proc sgplot data=sashelp.cars;
  where type='Sedan' or type='Sports';
  scatter x=horsepower y=mpg_city / transparency=0.9
       markerattrs=(symbol=circlefilled);
  reg x=horsepower y=mpg_city / nomarkers lineattrs=(thickness=2)
     cli clm degree=2;
  keylegend / location=inside position=topright across=1;
  xaxis grid;  yaxis grid;  run;
```

Figure 3.2.6: Grouped Fit Plot with Limits

This shows a grouped regression plot with confidence limits and 2nd order fit. CLMTRANSPARENCY is set on the REG statement.

To customize the legend, we have provided a name for the REG plot, and included it in the KEYLEGEND statement.

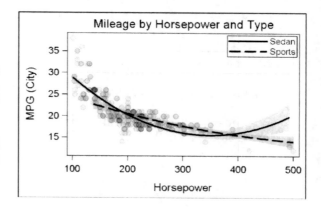

```
title 'Mileage by Horsepower and Type';
proc sgplot data=sashelp.cars;
  where type='Sedan' or type='Sports';
  scatter x=horsepower y=mpg_city / group=type transparency=0.9
       markerattrs=(symbol=circlefilled);
  reg x=horsepower y=mpg_city / group=type nomarkers name='a'
       lineattrs=(thickness=2) degree=2 clm clmtransparency=0.6;
  keylegend 'a' / location=inside position=topright across=1;
  xaxis grid;   yaxis grid;   run;
```

Figure 3.2.7: Histogram

This graph displays a basic distribution of mileage for all cars using a histogram.

The y-axis grid lines are enabled.
The x-axis label is suppressed.

By default the histogram shows a LINEAR axis, as shown here. To show a binned axis, use the SHOWBINS option.

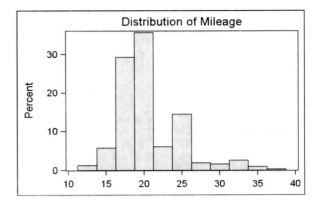

```
title 'Distribution of Mileage';
proc sgplot data=sashelp.cars;
  where type='Sedan' or type='Sports';
  histogram mpg_city;
  xaxis display=(nolabel);
  yaxis grid;
  run;
```

Figure 3.2.8: Histogram with Density Plots

This graph shows the distribution of horsepower for all cars, along with the Normal and Kernel density functions.

The legend is placed in the upper right corner of the graph.

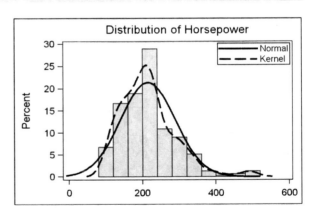

```
title 'Distribution of Horsepower';
proc sgplot data=sashelp.cars;
  where type='Sedan' or type='Sports';
  histogram horsepower;
  density horsepower;
  density horsepower / type=kernel;
  keylegend / location=inside position=topright across=1;
  xaxis display=(nolabel);
  yaxis grid;
  run;
```

Figure 3.2.9: Box Plot – Vertical

This graph displays the distribution of mileage for all cars by type using a box plot.

The x-axis label is suppressed since this information is included in the title already.

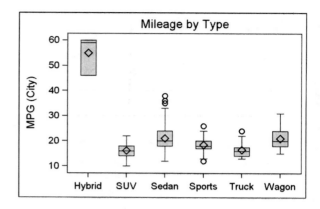

```
title 'Mileage by Type';
proc sgplot data=sashelp.cars;
  vbox mpg_city / category=type;
  xaxis display=(nolabel);
  yaxis grid;
  run;
```

Figure 3.2.10: Grouped Box Plot

This graph displays the distribution of mileage for vehicles grouped by origin using a horizontal box plot.

The line attributes are set to make all box outlines solid, and the legend is moved into the bottom right corner of the wall.

The x-axis label is suppressed.

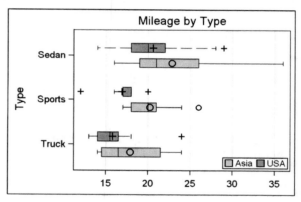

```
title 'Mileage by Type';
proc sgplot data=sashelp.cars;
  where (type='Sedan' or type='Sports' or type='Truck') and
        (origin='USA' or origin='Asia');
  hbox mpg_city / category=type group=origin groupdisplay=cluster
        lineattrs=(pattern=solid);
  keylegend / location=inside position=bottomright;
  xaxis grid display=(nolabel);
  run;
```

Figure 3.2.11: Interval Box Plot

This graph displays the distribution of horsepower by mileage on an interval axis.

The width of the boxes is based on the minimum interval between x-axis values.

Note: Box plot for interval data is a SAS 9.3 feature. The TYPE=interval option itself for XAXIS is available with SAS 9.2.

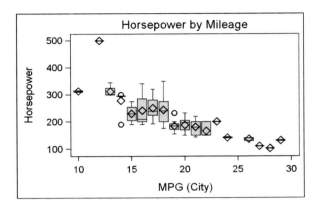

```
title 'Horsepower by Mileage';
proc sgplot data=sashelp.cars;
  where origin='USA';
  vbox Horsepower / category=mpg_city;
  xaxis type=linear;
  yaxis grid;
  run;
```

Figure 3.2.12: Series Plot

This graph shows a traditional stock plot of closing price by month using a series plot.

The x-axis label is suppressed since it is redundant.

A WHERE clause is used to display the data for only one value of the stock variable.

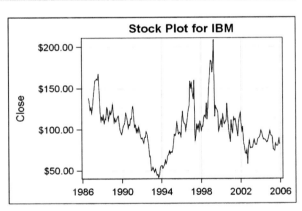

```
title 'Stock Plot for IBM';
proc sgplot data=sashelp.stocks;
  where stock = 'IBM';
  series x=date y=close;
  xaxis grid display=(nolabel);
  yaxis grid;
  run;
```

Figure 3.2.13: Grouped Series Plot

This graph shows a plot of closing stock price by month for three companies.

Patterned lines are hard to follow when the data fluctuates rapidly. This is a case where a color style would be more effective.

ANTIALIASMAX is set to 1000 to prevent disabling of antialiasing for a large number of observations.

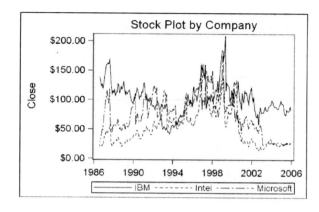

```
ods graphics / antialiasmax=1000;
title 'Stock Plot by Company';
proc sgplot data=sashelp.stocks;
  series x=date y=close / group=stock;
  keylegend / location=outside across=3;
  xaxis grid display=(nolabel);
  yaxis grid;
  run;
```

Figure 3.2.14: Multi-Response Series

This graph shows the efficacy of medication over time for three different treatment groups using multiple series plots. Each plot is labeled directly, making a legend unnecessary.

The procedure automatically assigns different visual attributes to each plot. By default, the time axis displays the year value only when needed.

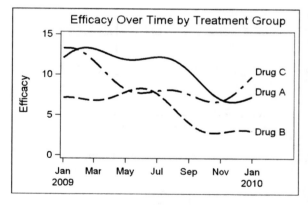

```
title 'Efficacy Over Time by Treatment Group';
proc sgplot data=sgbook.series noautolegend;
  series x=date y=a / lineattrs=(thickness=2) curvelabel='Drug A';
  series x=date y=b / lineattrs=(thickness=2) curvelabel='Drug B';
  series x=date y=c / lineattrs=(thickness=2) curvelabel='Drug C';
  xaxis display=(nolabel);
  yaxis min=0 max=15 integer grid label='Efficacy';
  run;
```

Figure 3.2.15: Grouped Step Plot

This graph shows a traditional stock plot of closing price by month for three companies using a grouped step plot.

This is similar to Figure 3.2.13, with use of a step plot instead of a series plot.

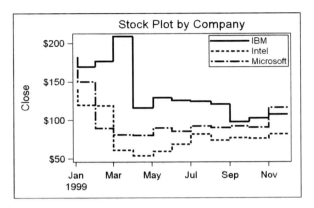

```
title 'Stock Plot by Company';
proc sgplot data=sashelp.stocks;
  where date > '01Jan99'd and date < '01Jan00'd;
  format close dollar5.0;
  step x=date y=close / group=stock lineattrs=(thickness=2);
  xaxis grid display=(nolabel);
  yaxis grid;
  keylegend / location=inside position=topright across=1;
  run;
```

Figure 3.2.16: Step Plot with Limits

This plot shows the predicted weight by height for students in a classroom using a step plot.

Confidence values are shown as Y error bars. LINEATTRS option is used to set the thickness of the step plot line, and ERRORBARATTRS is used to set the pattern for the error bars.

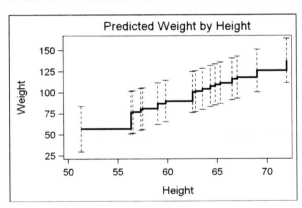

```
title 'Predicted Weight by Height';
proc sgplot data=sashelp.classfit noautolegend;
   step x=height y=predict / yerrorlower=lower yerrorupper=upper
        lineattrs=(thickness=2) errorbarattrs=(pattern=dash);
   xaxis grid;
   yaxis grid label='Weight';
   run;
```

Figure 3.2.17: Dot Plot with Limits

This graph displays the mileage for cars by type using a dot plot. The mean value is represented by the dot. The 95% confidence limits are displayed.

The legend is placed inside the data area at the bottom. To create some space for this, we have to set the OFFSETMAX=0.2. This is because the DOT plot sets REVERSE on the y-axis.

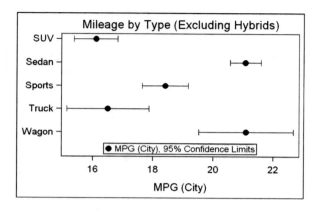

```
title 'Mileage by Type (Excluding Hybrids)';
proc sgplot data=sashelp.cars;
  where type ne 'Hybrid';
  dot type / response=mpg_city stat=mean limits=both;
  keylegend / location=inside position=bottom;
  xaxis grid;
  yaxis offsetmax=0.2 display=(nolabel);
run;
```

Figure 3.2.18: Vector Plot

This graph shows the mileage by horsepower for sedans made in USA as related to customer preferences of 20 MPG and 200 HP using a vector plot.

X and Y origins for the vector plot are set to constant values, so that all the vectors have the same starting point.

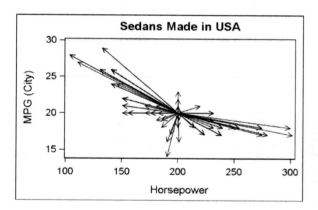

```
title 'Sedans Made in USA';
proc sgplot data=sashelp.cars;
  where type= 'Sedan' and origin='USA';
  vector x=horsepower y=mpg_city / xorigin=200 yorigin=20;
  xaxis grid; yaxis grid;
run;
```

Figure 3.2.19: Grouped Vector Plot

This graph shows the particle location and velocity using a vector plot. Direction (forward or backwards) is indicated by the group variable

Here, the X and Y origin roles are assigned columns from the dataset, so each vector has different start and end point.

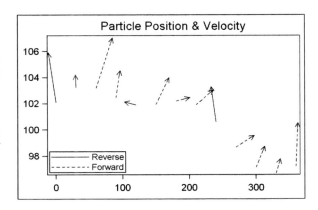

```
title 'Particle Position & Velocity';
proc sgplot data=sgbook.vectorGroup;
   vector x=x y=y / xorigin=xo yorigin=yo group=group lineattrs=(thickness=2);
   xaxis grid display=(nolabel);
   yaxis grid display=(nolabel);
   keylegend / location=inside position=bottomleft across=1;
   run;
```

Figure 3.2.20: Bar Chart

This is a basic bar chart of the mean mileage by country of origin.

Data labels are displayed using the DATALABEL option. A format is applied to the variable to reduce the label precision.

The x-axis label is suppressed as it is redundant.

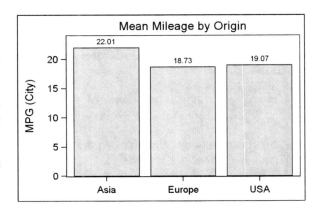

```
title 'Mean Mileage by Origin';
proc sgplot data=sashelp.cars;
   format mpg_city 5.2;
   vbar origin / response=mpg_city stat=mean datalabel;
   xaxis display=(nolabel);
   yaxis grid;
   run;
```

Figure 3.2.21: Grouped Bar Chart with Skins

This is a grouped bar chart of mileage by vehicle type with country of origin as the group variable.

GROUPDISPLAY=cluster is set to display adjacent bars for each group.

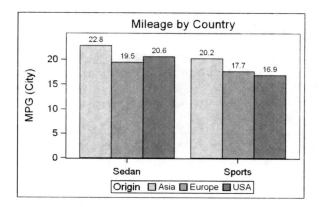

```
title 'Mileage by Country';
proc sgplot data=sashelp.cars;
  where type='Sedan' or type='Sports';
  format mpg_city 5.1;
  vbar type / response=mpg_city group=origin stat=mean
       groupdisplay=cluster datalabel;
  xaxis display=(nolabel);
  yaxis grid;
run;
```

Figure 3.2.22: Stacked Bar Chart with Skins

This is a grouped bar chart of sales by product with region as the group variable.

The group values are stacked by default. The x-axis label is suppressed.

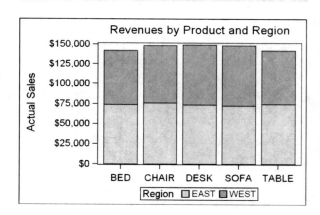

```
title 'Revenues by Product and Region';
proc sgplot data=sashelp.prdsale;
  format actual dollar8.0;
  vbar product / response=actual group=region;
  xaxis display=(nolabel);
  yaxis grid;
run;
```

Figure 3.2.23: Bar-Line Overlay

This is a basic bar-line overlay of two different measures by quarter.

The x-axis label is suppressed. The line thickness is set using the LINEATTRS option. Transparency is set to 30%.

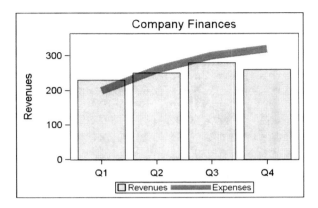

```
title 'Company Finances';
proc sgplot data=sgbook.quarter;
  vbar quarter / response=revenues nostatlabel;
  vline quarter / response=expenses nostatlabel
        lineattrs=(thickness=10) transparency=0.3;
  yaxis grid offsetmin=0;  xaxis display=(nolabel);
run;
```

Figure 3.2.24: Bar-Bar Overlay

This is a bar-bar overlay of two different measures by country. Bar width for the second bar chart is reduced.

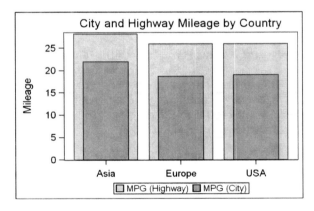

```
title 'City and Highway Mileage by Country';
proc sgplot data=sashelp.cars;
  vbar origin / response=mpg_highway stat=mean nostatlabel;
  vbar origin / response=mpg_city stat=mean
        nostatlabel barwidth=0.6;
  xaxis display=(nolabel);
  yaxis grid offsetmin=0 label='Mileage';
run;
```

Figure 3.2.25: Bar Chart with Multiple Responses

This graph shows two separate response variables plotted by country of origin. Here the two responses are displayed side by side using DISCRETEOFFSET and BARWIDTH options. The yaxis label is used.

This graph uses DATASKIN=Gloss to apply a decorative finish to the bars.

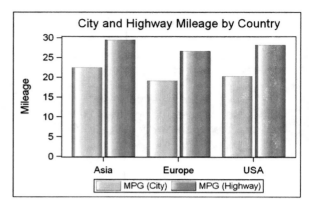

```
title 'City and Highway Mileage by Country';
proc sgplot data=sashelp.cars;
  where type='Sedan' or type='Sports';
  vbar origin / response=mpg_city stat=mean dataskin=gloss
       discreteoffset=-0.22 barwidth=0.4;
  vbar origin / response=mpg_highway stat=mean dataskin=gloss
       discreteoffset=+0.22 barwidth=0.4;
  xaxis display=(nolabel);
  yaxis label="Mileage";   run;
```

Figure 3.2.26: Bar Chart with Patterns

This graph shows a vertical bar chart of mileage by country of origin with vehicle type as group variable. The group values are stacked by default.

This graph uses the Journal2 style. The bars do not have fill colors, only fill patterns (9.3).

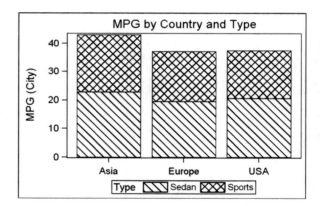

```
ods listing style=journal2;
title 'MPG by Country and Type';
proc sgplot data=sashelp.cars;
  where type='Sedan' or type='Sports';
  vbar origin / response=mpg_city stat=mean  group=type;
  xaxis display=(nolabel);
  run;
```

Figure 3.2.27: Grouped Bar Chart with Skins

This graph shows mileage by origin using a vertical bar chart with GROUP=type. GROUPDISPLAY=cluster is used to display adjacent bars for each group.

This graph uses DATASKIN=Sheen.

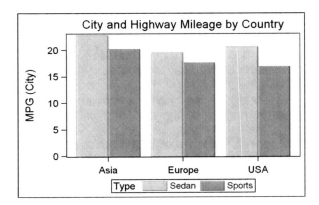

```
title 'City and Highway Mileage by Country';
proc sgplot data=sashelp.cars;
  where type='Sedan' or type='Sports';
  vbar origin / response=mpg_city stat=mean group=type
         groupdisplay=cluster dataskin=sheen transparency=0;
  xaxis display=(nolabel);
run;
```

Figure 3.2.28: Multiple Bar Chart with Patterns, Fill Colors, and Skins

This graph shows city and highway mileage by origin using adjacent bars with fill patterns, fill colors, and skins.

The Journal3 style uses color shades and fill patterns. The DISCRETEOFFSET and BARWIDTH options are used to position the two responses side by side.

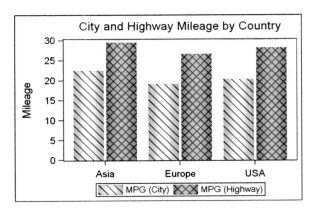

```
ods listing style=journal3;
title 'City and Highway Mileage by Country';
proc sgplot data=sashelp.cars;
  where type='Sedan' or type='Sports';
  vbar origin / response=mpg_city stat=mean barwidth=0.4
         dataskin=pressed discreteoffset=-0.22;
  vbar origin / response=mpg_highway stat=mean barwidth=0.4
         dataskin=pressed discreteoffset=+0.22;
  xaxis display=(nolabel);
  yaxis label='Mileage';   run;
```

Figure 3.2.29: Horizontal Bar Chart

This is a basic horizontal bar chart of mileage by car type, with data labels.

A horizontal bar chart is more suitable for larger number of bars and long tick values.

The x-axis and y-axis labels are suppressed as they are redundant.

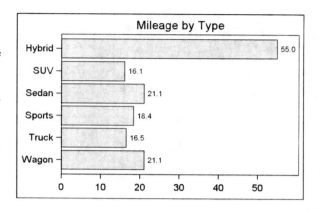

```
title 'Mileage by Type';
proc sgplot data=sashelp.cars;
  format mpg_city 5.1;
  hbar type / response=mpg_city stat=mean datalabel;
  xaxis grid display=(nolabel);
  yaxis display=(nolabel);
  run;
```

Figure 3.2.30: Horizontal Bars with Limits

This is a basic horizontal bar chart of mileage by car type, with limit bars by using the STAT and LIMIT options.

The y-axis label is disabled by using the DISPLAY option.

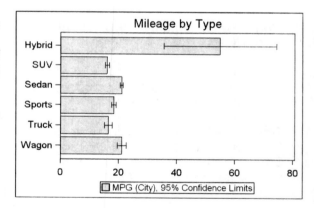

```
title 'Mileage by Type';
proc sgplot data=sashelp.cars;
  hbar type / response=mpg_city stat=mean limits=both;
  xaxis grid display=(nolabel);
  yaxis display=(nolabel);
  run;
```

Figure 3.2.31: HighLow Plot

This is a traditional stock plot of price by date, showing the high, low, open, and close prices using a HighLow plot with the OPEN and CLOSE options.

The default TYPE for the HighLow plot is 'Line'.

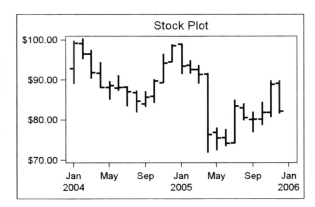

```
title 'Stock Plot';
proc sgplot data=sashelp.stocks;
   where stock='IBM' and date > '01Jan2004'd;
   highlow x=date high=high low=low / open=open close=close
           lineattrs=(thickness=2);
   xaxis grid display=(nolabel);
   yaxis grid display=(nolabel);
   run;
```

Figure 3.2.32: HighLow Plot

This is a plot showing duration of medications used for a clinical trial. The duration for each drug is plotted.

In this example, the TYPE for the HighLow plot is set to 'Bar'. The HIGHCAP and LOWCAP options are set to the 'cap' data variable to display bar end caps.

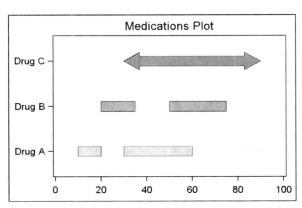

```
title 'Medications Plot';
proc sgplot data=sgbook.highlow noautolegend;
   highlow y=drug low=low high=high / group=drug
           lineattrs=(pattern=solid) highcap=cap lowcap=cap
           transparency=0.2 type=bar barwidth=0.4;
   yaxis   display=(nolabel) grid;
   xaxis   display=(nolabel) min=0 max=100 grid;
   run;
```

Figure 3.2.33: Bubble Plot

This is a grouped bubble plot, showing the response by size. Reference lines are used to demarcate the quadrants.

The max and min response values are mapped to max and min bubble sizes by radius.

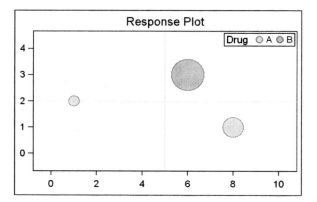

```
title 'Response Plot';
proc sgplot data=sgbook.bubble;
  refline 5 / axis=x transparency=0.7;
  refline 2 / axis=y transparency=0.7;
  bubble x=x y=y size=value / group=group transparency=0.4;
  keylegend / title='Drug' location=inside position=topright;
  yaxis   display=(nolabel) min=0 max=4;
  xaxis   display=(nolabel) min=0 max=10;
run;
```

Figure 3.2.34: Bubble Plots

This is a grouped bubble plot, showing the response by size. Here we are using unfilled bubbles. The groups are displayed using line patterns for the outline.

This uses the Journal style and NOFILL option. BRADIUSMAX is used to set the maximum size of the bubble in the graph to 50 pixels.

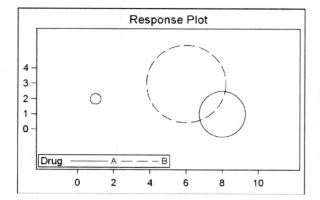

```
title 'Response Plot';
proc sgplot data=sgbook.bubble;
  bubble x=x y=y size=value / group=group bradiusmax=50 nofill;
  keylegend / title='Drug' location=inside position=bottomleft;
  yaxis   display=(nolabel) min=0 max=4;
  xaxis   display=(nolabel) min=0 max=10;
run;
```

3.3 Classification Panels

Figure 3.3.1: Bar Chart Panel

This is a classification panel using origin as the class variable.

Each cell contains a bar chart of mileage by type. Skins and transparencies can be combined to achieve different effects.

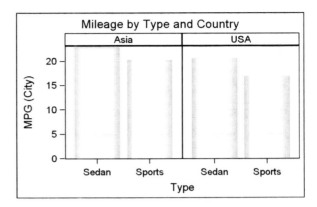

```
title 'Mileage by Type and Country';
proc sgpanel data=sashelp.cars ;
  where (type='Sedan' or type='Sports') and
        (origin='USA' or origin='Asia');
  panelby origin / novarname;
  vbar type / response=mpg_city stat=mean dataskin=gloss
              transparency=0.6;
  rowaxis grid;   run;
```

Figure 3.3.2: Box Plot Row Lattice

This is a classification panel using type as the row class variable. Each cell contains a box plot of mileage by origin.

This uses LAYOUT=ROWLATTICE to create multiple rows in a single column.

The COLAXIS statement is used to display the grid lines for the column axis.

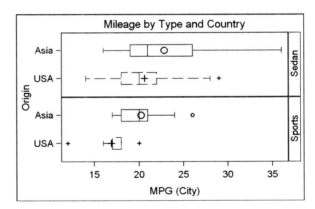

```
title 'Mileage by Type and Country';
proc sgpanel data=sashelp.cars noautolegend;
  where (type='Sedan' or type='Sports') and
        (origin='USA' or origin='Asia');
  panelby type / layout=rowlattice novarname;
  hbox mpg_city / category=origin group=origin;
  colaxis grid;
  run;
```

Figure 3.3.3: Histogram Lattice

This is a 2x2 lattice displaying the distribution of cholesterol by sex and blood pressure status using the HISTOGRAM and DENSITY statements.

The ROWAXIS and COLAXIS statements are used to display the grid lines.

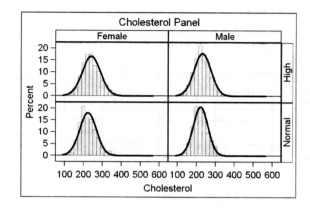

```
title 'Cholesterol Panel';
proc sgpanel data=sashelp.heart noautolegend;
  where bp_status='Normal' or bp_status='High';
  panelby sex bp_status / layout=lattice novarname;
  histogram cholesterol / transparency=0.5;
  density cholesterol;
  rowaxis grid;  colaxis grid;
run;
```

Figure 3.3.4: Scatter Panel

This graph has three class variables: type, origin, and drivetrain. Each cell has three headers, one for each class variable.

The header labels can take up a large portion of the cell space. Here, we have reduced the header label font size (in the style) to allow more space for the plot.

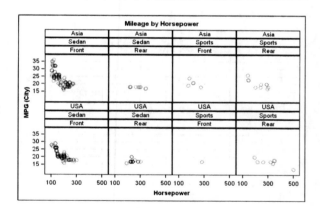

```
title 'Mileage by Horsepower';
proc sgpanel data=sashelp.cars ;
  where (type='Sedan' or type='Sports') and
        (origin='USA' or origin='Asia') and
        (drivetrain='Front' or drivetrain='Rear');
  panelby origin type drivetrain / novarname onepanel columns=4;
  scatter x=horsepower y=mpg_city / transparency=0.6;
  rowaxis grid;  colaxis grid;
  run;
```

3.4 Comparative and Matrix Graphs

Figure 3.4.1: Comparative Scatter Plot

This is a grid of scatter plots comparing systolic and diastolic blood pressure by cholesterol and weight.

The graph uses common external axes, so the data is uniform across rows and columns.

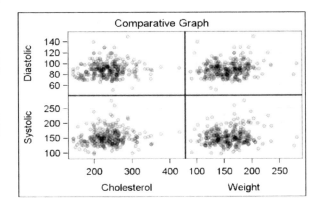

```
title 'Comparative Graph';
proc sgscatter data=sashelp.heart;
   where ageatstart > 58;
   compare x=(cholesterol weight) y=(diastolic systolic) /
           transparency=0.8 markerattrs=(symbol=circlefilled) grid;
   run;
```

Figure 3.4.2: Scatter Plot Matrix

This is a scatter plot matrix, showing the association between variables for patients over the age of 58.

Transparency can be used to see the distribution of the data. The diagonals display the distribution of each variable by itself.

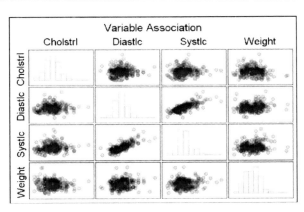

```
title 'Variable Association';
proc sgscatter data=sashelp.heart;
   label weight='Weight' cholesterol='Cholstrl'
         diastolic='Diastlc' systolic='Systlc';
   where ageatstart > 58;
   matrix cholesterol diastolic systolic weight / diagonal=(histogram)
         transparency=0.8 markerattrs=(symbol=circlefilled);
   run;
```

Chapter 4

Basic Plots

4.1 Introduction 61
4.2 SGPLOT Procedure 62
4.3 Plot Roles and Options 63
4.4 Scatter Plot 64
4.5 Scatter Plots with Data Labels 74
4.6 Series Plot 78
4.7 Step Plot 84
4.8 Band Plot 89
4.9 Needle Plot 94
4.10 Vector Plot 97
4.11 VBarParm and HBarParm Plots (9.3) 100
4.12 Bubble Plot (9.3) 105
4.13 HighLow Plot (9.3) 108
4.14 Reference Lines 112
4.15 Parametric Line Plot (9.3) 115
4.16 Waterfall Chart (9.3) 117
4.17 Combining the Plots 120

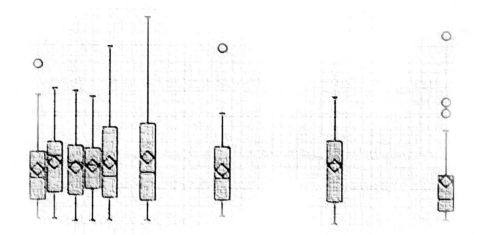

*"If you can't explain it simply,
you don't understand it well enough."
~ Albert Einstein*

Chapter 4: Basic Plots

In this chapter, we examine the basic plots for SGPLOT and SGPANEL procedures. These plots have an extensive set of options that will be covered in detail in this chapter. Plots within this group can be mixed and matched with each other in a myriad of ways to build complex graphs as shown in Figure 3.2.5.

4.1 Introduction

Basic plots used in both the SGPLOT and SGPANEL procedure are the workhorse plots for visualization of raw data, or of summary statistics that have been computed prior to creation of the graph. These plots do not process the data in any way, but they plot the raw data values in the graph. The plots included in this group are:

- SCATTER
- NEEDLE
- BUBBLE
- WATERFALL
- SERIES
- VECTOR
- HIGHLOW
- STEP
- VBARPARM
- REFLINE
- BAND
- HBARPARM
- LINEPARM

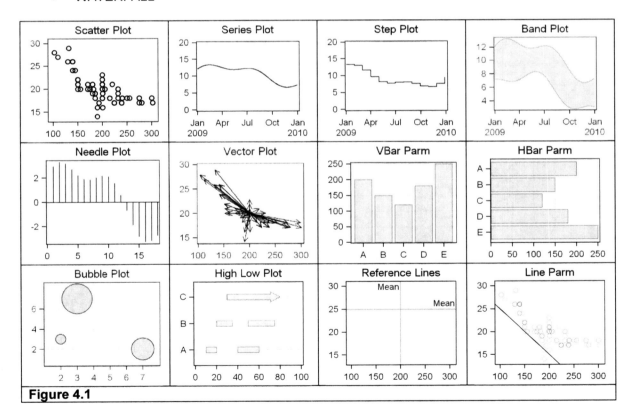

Figure 4.1

4.2 SGPLOT Procedure

The SGPLOT procedure supports the following syntax:

```
proc sgplot data=datasetname <options>;
```

Options:

CYCLEATTRS	*boolean*	Cycle through style elements for each plot
DATA	*=sas-data-set*	Optional data set
DATTRMAP	*=sas-data-set*	Data set defining an attribute map
DESCRIPTION	*=string*	Description string
NOAUTOLEGEND	*boolean*	Do not show the automatic legend
NOCYCLEATTRS	*boolean*	Do not cycle style elements for each plot
PAD	*=value*	Padding around the outside of the graph
SGANNO	*=sas-data-set*	Data set containing the annotations
TMPLOUT	*=string*	File name for generated graph template code
UNIFORM	*=keyword*	Uniform axis and legends across BY variables

The procedure supports the following statements:
- Basic Plots: BAND, NEEDLE, SCATTER, STEP, SERIES, VECTOR, BUBBLE, HIGHLOW, HBARPARM, VBARPARM
- Fit Plots: LOESS, PBSPLINE, REG, LINEPARM
- Distribution Plots: DENSITY, ELLIPSE HISTOGRAM, HBOX, VBOX
- Categorical Plots: DOT, HBAR, HLINE, VBAR, VLINE, WATERFALL
- Other: KEYLEGEND, INSET, REFLINE, XAXIS, YAXIS, X2AXIS, Y2AXIS

The procedure has the following syntax:

```
PROC SGPLOT < DATA= data-set > < options >;
  plot-statement(s) required-parameters < / options >;
  < refline-statement(s) >;
  < inset-statement(s) >;
  < keylegend-statement(s) >;
  < axis-statement(s) >;
RUN;
```

As seen in the examples in Chapter 3, one or more plot statements can be provided that will work together to create the graph. See the table of permissible combinations shown in section 2.4.

4.3 Plot Roles and Options

All of the plot statements have the following syntax:

```
PlotName required-data-roles </ options>;
```

All plot statements have required and optional parameters necessary to create the plot. These parameters fall in the following broad categories:

1. required data roles
2. optional data roles
3. options
4. common options

Categories 1 - 3 above include role and option names that are specific to each plot. These role and option names are used consistently and only as needed. We will discuss these with each specific plot statement.

Category 4 includes the set of options that are common to all plot statements as shown in the table below. These options work in the same way across all plot statements. Some plots may not support one or more of these common options.

Common Options:

ATTRID	=string	The attribute map ID (See Chapter 9).
CLUSTERWIDTH	=value	Fraction of the mid-point spacing to be used for drawing the group cluster.
DISCRETEOFFSET	=value	Fractional shift within midpoint spacing.
GROUPDISPLAY	=keyword	OVELAY | CLUSTER. Default value varies.
GROUPORDER	=keyword	DATA | ASCENDING | DESCENDING
LEGENDLABEL	=string	The label that appears in the legend to represent this (non-group) plot.
NAME	=string	Specifies a name for this statement. Other statements, such as KEYLEGEND, can refer to a plot by its name.
TRANSPARENCY	=value	Specifies transparency for the visual elements.
URL	=string-column	URL link when used in HTML output.

Common Boolean Options:

NOMISSINGGROUP	Missing values are not displayed as a separate group.
X2AXIS	Assigns X values to the X2 (top) axis.
Y2AXIS	Assigns Y values to the Y2 (right) axis.

These common options are available for each plot statement, and can be used exactly as described above. To avoid duplication, we will not list the above common options for the individual plots that are discussed in this chapter.

4.4 Scatter Plot

The basic scatter plot is a key plot type for visualization of the raw data. The syntax is:

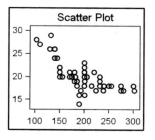

```
scatter x=column y=column </ options>;
```

A marker is displayed at each (x, y) location. Group, Data Labels, Marker Character, and X and Y Error Bars are supported. Discrete axes can have cluster grouping and discrete offsets.

Required Data Roles:

X	=column	Position along the horizontal axis
Y	=column	Position along the vertical axis

Optional Data Roles:

DATALABEL	=column	Label String for each observation
FREQ	=num-column	Frequency count for each observation
GROUP	=column	Classification variable for each observation
MARKERCHAR	=column	String to be shown for the observation
URL	=column	URL for each observation
XERRORLOWER	=num-column	Lower value for X Error Bar
XERRORUPPER	=num-column	Upper value for X Error Bar
YERRORLOWER	=num-column	Lower value for Y Error Bar
YERRORUPPER	=num-column	Upper value for Y Error Bar

Appearance Options:

ERRORBARATTRS	=line-attrs	Appearance attributes for error bars
MARKERATTRS	=marker-attrs	Appearance attributes for markers
MARKERCHARATTRS	=text-attrs	Appearance attributes for char markers

Boolean Options:

DATALABEL	Display Y value as data label

Standard common options are supported
ATTRID, CLUSTERWIDTH, DISCRETEOFFSET, GROUPDISPLAY, GROUPORDER, LEGENDLABEL, NAME, NOMISSINGGROUP, TRANSPARENCY, URL, X2AXIS, Y2AXIS

Related Style Elements: The GraphDataDefault style element is used to draw the plot elements. When a GROUP variable is present, or when multiple plot statements are overlaid, the GraphData1-GraphData12 elements are used, one for each group or plot statement.

Figure 4.4.1: Basic Scatter Plot

This is a basic scatter plot of mileage by horsepower. Marker attributes are customized to use the 2nd data element with filled markers.

Marker transparency is set to 80%. This reduces clutter and helps to view the regions of higher density.

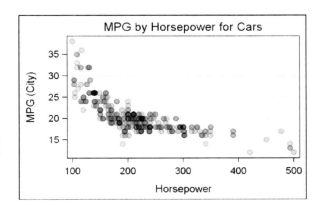

```
title 'MPG by Horsepower for Cars';
proc sgplot data=sashelp.cars;
  where type='Sedan' or type='Sports';
  scatter x=horsepower y=mpg_city /
       markerattrs=graphdata2(symbol=circlefilled) transparency=0.8;
  xaxis grid;
  yaxis grid;
  run;
```

Figure 4.4.2: Grouped Scatter Plot

When a GROUP variable is set, the marker attributes of each group are derived from the GraphData (1-N) style elements.

An automatic legend is created for group values. We have used the KEYLEGEND statement to position the legend inside the data area in the upper right corner.

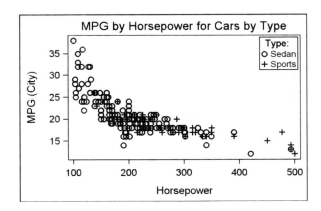

```
title 'MPG by Horsepower for Cars by Type';
proc sgplot data=sashelp.cars;
  where type='Sedan' or type='Sports';
  scatter x=horsepower y=mpg_city / group=type;
  keylegend / title='Type: ' location=inside across=1;
  xaxis grid;
  yaxis grid;
  run;
```

Figure 4.4.3: Two Scatter Plot Overlay

This is an overlay of two scatter plots. The procedure automatically assigns each plot with different visual attributes.

The legend is automatically generated. The KEYLEGEND statement is used here to place the legend in the upper right corner.

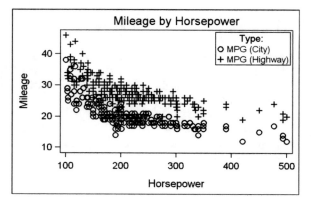

```
title 'Mileage by Horsepower';
proc sgplot data=sashelp.cars;
  where type='Sedan' or type='Sports';
  scatter x=horsepower y=mpg_city;
  scatter x=horsepower y=mpg_highway;
  keylegend / title='Type: ' location=inside across=1;
  xaxis grid;
  yaxis grid label='Mileage';   run;
```

Figure 4.4.4: Two Scatter Plot with Custom Attributes

This graph uses overlay of two scatter plots. The MARKERATTRS option is used to specify the marker attributes for each plot.

Legend is positioned in the upper right corner of the data area by using the KEYLEGEND statement.

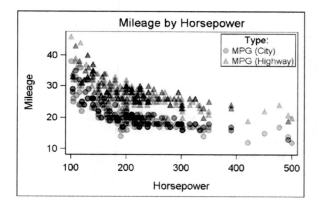

```
title 'Mileage by Horsepower';
proc sgplot data=sashelp.cars;
  where type='Sedan' or type='Sports';
  scatter x=horsepower y=mpg_city / transparency=0.7
          markerattrs=(symbol=circlefilled);
  scatter x=horsepower y=mpg_highway / transparency=0.7
          markerattrs=(symbol=trianglefilled);
  keylegend / title='Type: ' location=inside across=1;
  yaxis grid label='Mileage';
  xaxis grid;   run;
```

Figure 4.4.5: Scatter with Y Upper Error

This is a scatter plot with Y upper error limit.

The MARKERATTRS option is used to set the marker attributes. The x- and y-axis gridlines are enabled.

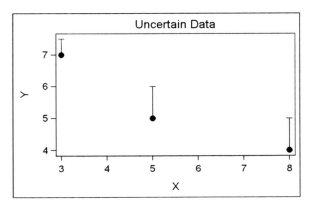

```
 title 'Uncertain Data';
proc sgplot data=sgbook.limits;
   scatter x=x y=y / yerrorupper=yh markerattrs=(symbol=circlefilled);
   xaxis grid;
   yaxis grid;
   run;
```

Figure 4.4.6: Scatter with Y Lower and Upper Error

This is a scatter plot with Y upper and lower error limits.

The MARKERATTRS option is used to set the marker attributes. The x- and y-axis gridlines are enabled.

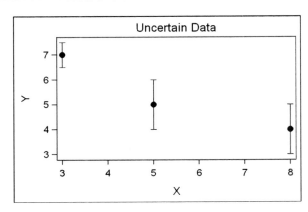

```
title 'Uncertain Data';
proc sgplot data=sgbook.limits noautolegend;
   scatter x=x y=y / yerrorlower=yl yerrorupper=yh
           markerattrs=(symbol=circlefilled);
   xaxis grid;
   yaxis grid;
   run;
```

Figure 4.4.7: Scatter with X Lower and Upper Error

This scatter plot displays X lower and upper error limits.

We have created filled markers with outline by overlaying two scatter plots. The NOAUTOLEGEND and NOCYCLEATTRS options are used to switch off automatic creation of the legend and cycling of the attributes.

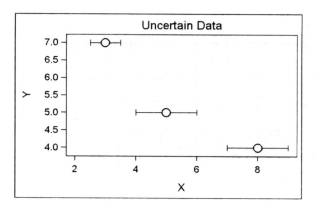

```
title 'Uncertain Data';
proc sgplot data=sgbook.limits noautolegend nocycleattrs;
   scatter x=x y=y / xerrorlower=xl xerrorupper=xh
           markerattrs=(size=11);
   scatter x=x y=y / markerattrs=(symbol=circlefilled color=white);
   xaxis grid;  yaxis grid;
   run;
```

Figure 4.4.8: Scatter with X and Y Limits

This scatter plot displays both lower and upper limits for X and Y.

We have created filled markers with outline by overlaying two scatter plots. The NOAUTOLEGEND and NOCYCLEATTRS options are used to switch off automatic creation of the legend and cycling of the attributes.

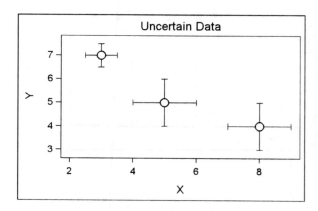

```
title 'Uncertain Data';
proc sgplot data=sgbook.limits noautolegend nocycleattrs;
   scatter x=x y=y / xerrorlower=xl xerrorupper=xh
           yerrorlower=yl yerrorupper=yh markerattrs=(size=11);
   scatter x=x y=y / markerattrs=( symbol=circlefilled color=white);
   xaxis grid;  yaxis grid;
   run;
```

Figure 4.4.9: Grouped Scatter with X and Y Limits

This shows a grouped scatter plot with X and Y limits.

To create the gap between the error bars and the markers, we have used multiple scatter plot statement to create this effect. This makes the marker shape more clearly visible.

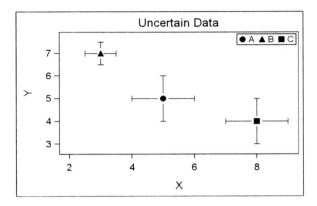

```
title 'Uncertain Data';
proc sgplot data=sgbook.limits nocycleattrs;
   scatter x=x y=y / group=group name='a'
           xerrorlower=xl xerrorupper=xh yerrorlower=yl yerrorupper=yh;
   scatter x=x y=y / group=group markerattrs=(size=13 color=white);
   scatter x=x y=y / group=group;
   keylegend 'a' / location=inside position=topright;
   xaxis grid;
   yaxis grid;
   run;
```

Figure 4.4.10: Scatter with X and Y Limits

This shows a grouped scatter plot with X and Y limits.

This graph uses a custom style that uses "filled" markers. Error bar attributes are set to a dash pattern and thicker line. The end cap size is tied to line thickness.

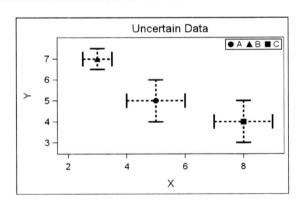

```
title 'Uncertain Data';
proc sgplot data=sgbook.limits;
   scatter x=x y=y / xerrorlower=xl xerrorupper=xh
           yerrorlower=yl yerrorupper=yh group=group
           errorBarAttrs=(thickness=2 pattern=dash);
   keylegend / location=inside position=topright;
   xaxis grid;
   yaxis grid;
   run;
```

Figure 4.4.11: Scatter Plot with Y2 Axis

This is a basic scatter plot with y-axis displayed on the right side (Y2).

We are using the custom style that uses filled markers. Transparency is set for a better view of the density of the observations.

Note the position of the y-axis as indicated by the arrow.

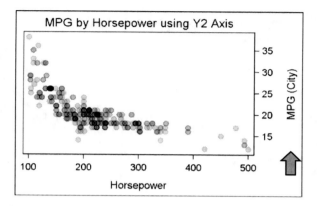

```
title 'MPG by Horsepower using Y2 Axis';
proc sgplot data=ssashelp.cars;
  where type='Sedan' or type='Sports';
  scatter x=horsepower y=mpg_city / transparency=0.8 Y2axis
          markerattrs=(symbol=circlefilled);
  xaxis grid;
  y2axis grid;
  run;
```

Figure 4.4.12: Scatter Plot with X2 and Y2 Axes

This is a basic scatter plot with x-axis displayed at the top (X2) and y-axis displayed on the right (Y2).

We are using the custom style that uses filled markers. Markers are transparent.

Note the positions of the axes as indicated by the arrows.

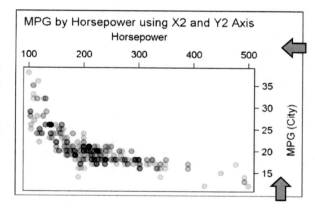

```
title 'MPG by Horsepower using X2 and Y2 Axis';
proc sgplot data=sashelp.cars;
  where type='Sedan' or type='Sports';
  scatter x=horsepower y=mpg_city / transparency=0.8 x2axis Y2axis
          markerattrs=(symbol=circlefilled);
  x2axis grid;
  y2axis grid;
  run;
```

Figure 4.4.13: Scatter with Reflected Ticks on Y2

This is a basic scatter plot with reflected ticks on the Y2 axis.

We are using a custom style with filled markers. This graph uses the GraphData3 style element that has square, filled symbols.

The ref ticks are indicated by the arrow.

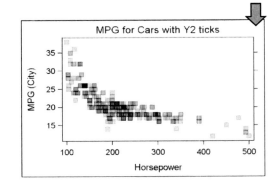

```
title 'MPG for Cars with Y2 ticks';
proc sgplot data=sashelp.cars;
  where type='Sedan' or type='Sports';
  scatter x=horsepower y=mpg_city / markerattrs=(symbol=squarefilled)
          transparency=0.8;
  xaxis grid;
  yaxis grid refticks;
  run;
```

Figure 4.4.14: Scatter with X, Y, and Y2 Axis

Low and high average temperatures by month are plotted using two scatter plots. Temperature in Fahrenheit is shown on the y-axis and in Celsius on the Y2 axis.

The title uses unicode text. XAXIS OFFSETMIN is used to leave extra space for the REFLINE label. The LEGENDLABEL option is used to name each plot in the legend.

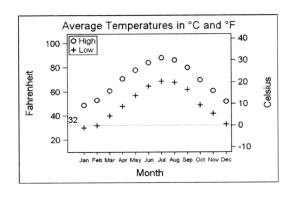

```
ods escapechar='^';
title "Average Temperatures in ^{unicode '00b0'x}C and ^{unicode '00b0'x}F";
proc sgplot data=sgbook.weather;
  refline 32 / label='32' labelloc=inside labelpos=min;
  scatter x=month y=high / legendlabel='High';
  scatter x=month y=lowc / y2axis legendlabel='Low';
  xaxis grid valueattrs=(size=6) offsetmin=0.1;
  yaxis grid min=14 max=104;
  y2axis min=-10 max=40;
  keylegend / location=inside position=topleft across=1;
  run;
```

Figure 4.4.15: Scatter with Groups on Discrete Axis

Here is a scatter plot of MPG by origin, with GROUP=Type. The x-axis variable is discrete. All markers are overlaid at the (x, y) location, resulting in a dense plot.

It is not easy to distinguish the different groups. A color style would have helped a bit. X-axis offsets are used to move the markers inwards.

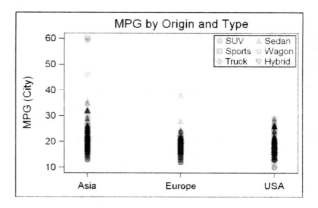

```
title "MPG by Origin and Type";
proc sgplot data=sashelp.cars;
  scatter x=origin y=mpg_city / group=type transparency=0.8;
  xaxis display=(nolabel) offsetmin=0.1 offsetmax=0.1;
  yaxis grid;
  keylegend / location=inside position=topright across=2;
  run;
```

Figure 4.4.16: Scatter with Clustered Groups on Discrete Axis

In this graph, we have the same data as in Figure 4.4.15, but we are using GROUPDISPLAY=Cluster.

When the x-axis variable is discrete, the GROUPDISPLAY option creates clusters for each group around the mid-point value, just like a bar chart.

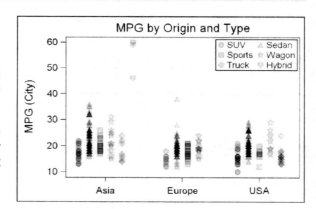

```
title "MPG by Origin and Type";
proc sgplot data=sashelp.cars;
  scatter x=origin y=mpg_city / group=type groupdisplay=cluster
          transparency=0.8;
  xaxis display=(nolabel);
  yaxis grid;
  keylegend / location=inside position=topright across=2;
  run;
```

Figure 4.4.17: Scatter Overlay

This graph is an overlay of two scatter plots of city and highway mileage by country of origin. The markers for each are overlaid, resulting in a cluttered plot.

To leave more room on the outside of the x-axis, x-axis offsets are used.

See Figure 4.4.18 below.

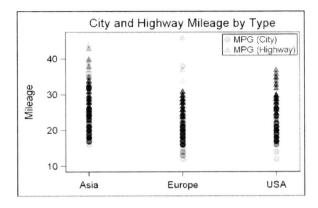

```
title "City and Highway Mileage by Type";
proc sgplot data=sashelp.cars;
   where type='Sedan' or type='Sports';
   scatter x=origin y=mpg_city / transparency=0.8;
   scatter x=origin y=mpg_highway / transparency=0.8;
   xaxis display=(nolabel) offsetmin=0.1 offsetmax=0.1;
   yaxis grid label='Mileage';
   keylegend / location=inside position=topright across=1;
   run;
```

Figure 4.4.18: Scatter Overlay with Discrete Offset

This graph has the same data as for Figure 4.4.17. Here, we have offset each scatter plot a little bit using the DISCRETEOFFSET option.

This creates a separation between the different Y values, making the graph more readable.

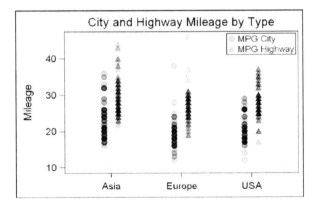

```
title "City and Highway Mileage by Type";
proc sgplot data=sashelp.cars;
   where type='Sedan' or type='Sports';
   scatter x=origin y=mpg_city / transparency=0.8 discreteoffset=-0.1;
   scatter x=origin y=mpg_highway / transparency=0.8 discreteoffset=+0.1;
   xaxis display=(nolabel) offsetmin=0.2 offsetmax=0.2;
   yaxis grid label='Mileage';
   keylegend / location=inside position=topright across=1;
   run;
```

4.5 Scatter Plots with Data Labels

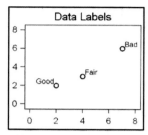

The SCATTER statement supports two ways to display data values as character strings:

- **MARKERCHAR = column**
- **DATALABEL < = column >**

4.5.1 DATALABEL Option

The scatter plot displays a marker at the (x, y) location for each observation. The SCATTER statement also supports display of data labels with each marker using the DATALABEL option. In this case, the associated data label is displayed near the marker:

- DATALABEL option displays the Y variable value.
- DATALABEL=column displays the values from the specified column.

By default, a data label collision avoidance algorithm is used to minimize the collisions between labels and other markers in the plot. Plots with data labels can get busy very quickly. When using data labels we recommend the following:

- Use short data labels.
- Restrict the number of markers.
- For moderately large number of markers, restrict the number of data labels. This can be done by selectively setting data labels to missing for some of the observations.

The data label collision avoidance algorithm attempts to move the labels to avoid collision with markers or other labels. This can sometimes move a label away from the marker. One technique to retain context is to use group colors as shown later.

4.5.2 MARKERCHAR Option

Often it is desirable to display a character string at the (x, y) position for each observation instead of a marker. The MARKERCHAR=column option draws a character string for the value from the column for each observation at the (x, y) location in the plot area.

Figure 4.5.5 extends the example in Figure 4.4.14 to display the temperature value for each observation. This makes it easier to see both the trend and the actual value.

Figure 4.5.1: Scatter with Data Labels

In this scatter plot we have used the DATALABEL option to display the data labels for Asian sedans with MPG > 27.

There are only 16 observations in this graph. However, due to relatively long labels, the graph quickly becomes cluttered. If all Asian sedans were displayed, the graph would be unreadable.

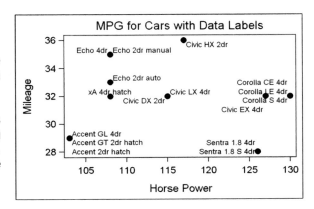

```
title 'MPG for Cars with Data Labels';
proc sgplot data=sgbook.cars3;
   where type='Sedan' and origin='Asia'  and mpgc > 27;
   scatter x=hp y=mpg / datalabel=model markerattrs=(symbol=circlefilled);
   xaxis grid;
   yaxis grid;
   run;
```

Figure 4.5.2: Scatter with Data Labels

In this scatter plot with data labels, we display only the observations for Asian sedans with MPG > 32.

When using data labels, it is generally better to reduce the displayed data to avoid a cluttered graph.

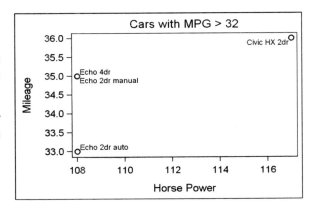

```
title 'Cars with MPG > 32';
proc sgplot data=sgbook.cars3 noautolegend;
   where type='Sedan' and origin='Asia'  and mpgc > 32;
   scatter x=hp y=mpg / datalabel=model;
   xaxis grid;
   yaxis grid;
   run;
```

Figure 4.5.3: Scatter with Short Labels

This is a basic scatter plot with short data labels for Asian sedans with MPG > 30.

In this example, we have shortened the data labels to 10 characters. Here we have 10 observations. Both readability and context are maintained by using shorter labels. Using the GROUP variable would help if we were using a color style.

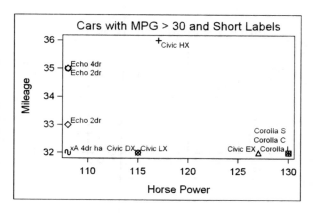

```
title 'Cars with MPG > 30 and Short Labels';
proc sgplot data=sgbook.cars3 noautolegend;
   where type='Sedan' and origin='Asia' and mpgc > 30;
   scatter x=hp y=mpg / datalabel=model2 group=model2;
   xaxis grid;
   yaxis grid;
   run;
```

Figure 4.5.4: Scatter with Reduced Number of Data Labels

This a scatter plot for all cars, showing data labels only for high-mileage or high-horsepower cars.

This example has 428 observations. However, we have modified the data set to display labels only for the exceptional cars by setting the other labels to missing.

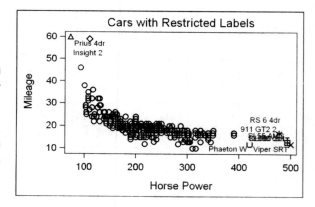

```
title 'Cars with Restricted Labels';
proc sgplot data=sgbook.cars4 noautolegend;
   scatter x=hp y=mpg / datalabel=model3 group=model3;
   xaxis grid;
   yaxis grid;
   run;
```

Figure 4.5.5: Scatter with Marker Characters

The MARKERCHAR option can be used to display a character string for the data in the column at the (x, y) location in the plot. Here we have displayed the high and low temperature values for each observation within the marker.

A Refline for y=32 is displayed.

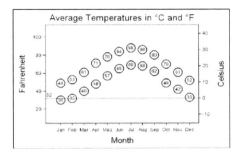

```
title "Average Temperatures in ^{unicode '00b0'x}C and ^{unicode '00b0'x}F";
proc sgplot data=sgbook.weather noautolegend;
  refline 32 / label='32' labelloc=inside labelpos=min;
  scatter x=month y=high / markerattrs=(symbol=circle size=15);
  scatter x=month y=high / markerchar=high;
  scatter x=month y=lowc / markerattrs=(symbol=circle size=15) y2axis;
  scatter x=month y=low / markerchar=low;
  xaxis grid offsetmin=0.1;
  yaxis grid min=14 max=104;
  y2axis min=-10 max=40;   run;
```

Figure 4.5.6: Scatter with Axis Aligned Statistics Table

The MARKERCHAR option is used to create a table of statistics that is aligned with a categorical y-axis.

REFLINE statements with DISCRETEOFFSET are used to create the vertical column borders. X2 axis is used with DISPLAY option set to simulate column headers.

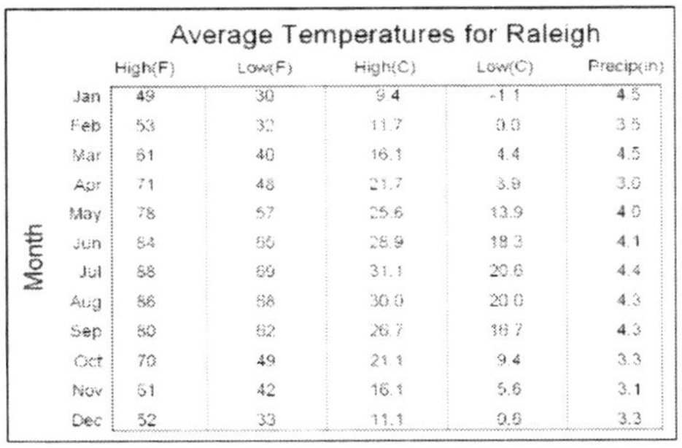

```
title "Average Temperatures for Raleigh";
proc sgplot data=sgbook.weatherTable noautolegend;
  scatter y=month x=x1 / markerchar=high x2axis;
  scatter y=month x=x2 / markerchar=low x2axis;
  scatter y=month x=x3 / markerchar=highc x2axis;
  scatter y=month x=x4 / markerchar=lowc x2axis;
  scatter y=month x=x5 / markerchar=precip x2axis;
  refline 'High(F)' 'Low(F)' 'High(C)' 'Low(C)' / axis=x2 discreteoffset=0.5;
  x2axis display=(noticks nolabel) valueattrs=(size=7);
  yaxis display=(noticks) reverse valueattrs=(size=7);
  run;
```

4.6 Series Plot

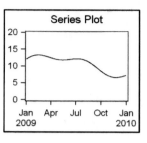

The basic series plot is a key plot type for visualization of raw data. This plot does not summarize the data. The syntax is:

```
series x=column y=column </ options>;
```

A series is drawn through each (x, y) location in data order. The GROUP role is used to create multiple series plots, one for each unique group value.

Required Data Roles:

X	=column	Position along the horizontal axis
Y	=column	Position along the vertical axis

Optional Data Roles:

DATALABEL	=column	Label String for each observation
GROUP	=column	Classification for multiple series
URL	=column	URL for each observation

Appearance Options:

LINEATTRS	=line-attrs	Appearance attributes for lines
MARKERATTRS	=marker-attrs	Appearance attributes for markers

Other Options:

CURVELABEL	=string	Curve label for non-grouped case
CURVELABELLOCATION	=keyword	Curve label location (INSIDE | OUTSIDE)
CURVELABELPOSITION	=keyword	Curve label position (TOP, BOTTOM, ...)

Boolean Options:

BREAK	Break the series plot at missing values
CURVELABEL	Display curve labels from group role
DATALABEL	Display Y value as data label
MARKERS	Display the markers

Standard common options are supported

ATTRID, CLUSTERWIDTH, DISCRETEOFFSET, GROUPDISPLAY, GROUPORDER, LEGENDLABEL, NAME, NOMISSINGGROUP, TRANSPARENCY, URL, X2AXIS, Y2AXIS

Related Style Elements: The GraphDataDefault style element is used to draw the plot elements. When a GROUP variable is present, or when multiple plot statements are overlaid, the GraphData1-GraphData12 elements are used, one for each group or plot statement.

Figure 4.6.1: Basic Series Plot

This graph shows patient response to a drug over time.

The x-axis label is suppressed, and YAXIS MIN is set to zero.

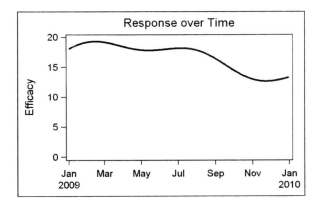

```
title 'Response over Time';
proc sgplot data=sgbook.seriesGroup;
   where drug='Drug A';
   series x=date y=val / lineattrs=(thickness=2);
   xaxis display=(nolabel);
   yaxis min=0 integer grid label='Efficacy';
   run;
```

Figure 4.6.2: Grouped Series Plot with Legend

This graph shows response over time by drug. A legend is generated by default. We have used the KEYLEGEND statement to suppress the legend title.

Since the number of observations in the data exceeds the default limit for anti-aliasing, set ANTIALIASMAX=1100 to restore anti-aliasing.

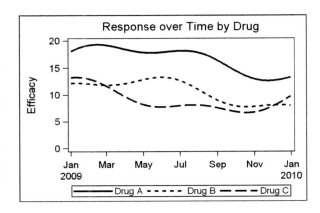

```
ods graphics / reset antialiasmax=1100;
title 'Response over Time by Drug';
proc sgplot data=sgbook.seriesGroup;
   series x=date y=val / group=drug lineattrs=(thickness=2);
   xaxis display=(nolabel);
   yaxis min=0 integer grid label='Efficacy';
   keylegend / title='' valueattrs=(size=7);
   run;
```

Figure 4.6.3: Grouped Series Plot with Curve Labels

In this graph the KEYLEGEND statement is used to move the legend closer to the data. This reduces the amount of eye movement needed to decode which series represents which drug.

See Figure 4.6.4 for an alternative arrangement.

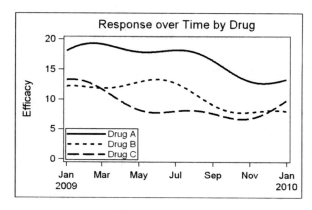

```
title 'Response over Time by Drug';
proc sgplot data=sgbook.seriesGroup;
  series x=date y=val / group=drug lineattrs=(thickness=2);
  xaxis display=(nolabel);
  yaxis min=0 integer grid label='Efficacy';
  keylegend / title='' location=inside position=bottomleft across=1;
run;
```

Figure 4.6.4: Grouped Series Plot with Markers

In this graph the legend is eliminated by labeling each series with the name of the drug. This eliminates eye movement required to decode the information, thus increasing the effectiveness of the graph.

The procedure detects the presence of the CURVELABEL option, and suppresses the legend automatically.

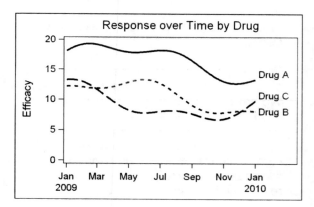

```
title 'Response over Time by Drug';
proc sgplot data=sgbook.seriesGroup;
  series x=date y=val / group=drug
         lineattrs=(thickness=2) curvelabel;
  xaxis display=(nolabel);
  yaxis min=0 integer grid label='Efficacy';
run;
```

Figure 4.6.5: Grouped Series Plot with Markers

In this graph we have displayed only some of the markers. This is done using an overlaid scatter plot with the FREQ= option. With this option, a marker is plotted only when the value of the FREQ variable is > 0.

We did this instead of using the markers option on the series plot as that would display a marker at each point on the line.

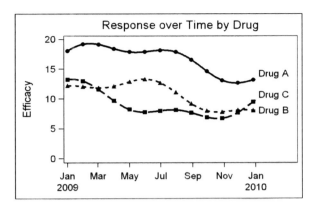

```
title 'Response over Time by Drug';
proc sgplot data=sgbook.seriesGroup noautolegend;
   series x=date y=val / group=drug
          lineattrs=(thickness=2) curvelabel;
   scatter x=date y=val / group=drug freq=freq markerattrs=(size=5);
   xaxis display=(nolabel);
   yaxis min=0 integer grid label='Efficacy';
run;
```

Figure 4.6.6: Series Plot with Data Labels

In this graph, only some data labels are displayed using an overlaid scatter plot. We could display data labels using the DATALABEL option of the series plot. However, that would display a label for every observation, cluttering the plot.

On the scatter plot, we use Y=LabelA which has non-missing values for only every 50th obs, thus reducing the markers and labels.

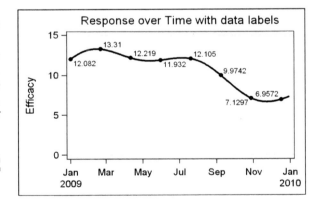

```
title 'Response over Time with data labels';
proc sgplot data=sgbook.series noautolegend;
   series x=date y=a / lineattrs=(thickness=2) datalabel=labela;
   scatter x=date y=labela / markerattrs=(size=5);
   xaxis display=(nolabel);
   yaxis min=0 max=15 integer grid label='Efficacy';
run;
```

Figure 4.6.7: Series Plot with Breaks

In this graph, some values of the response variable are missing. By default, the series plot would be drawn with a straight line over the missing data.

Using the BREAK option causes the series plot to be drawn with a break at the missing data.

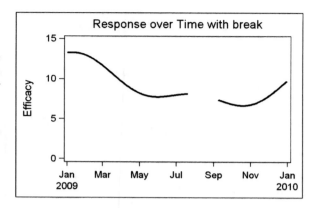

```
title 'Response over Time with break';
proc sgplot data=sgbook.seriesBreak;
   series x=date y=c / lineattrs=(thickness=2) break;
   xaxis display=(nolabel);
   yaxis min=0 max=15 integer grid label='Efficacy';
run;
```

Figure 4.6.8: Series Plot Overlay with Discrete Offset

This graph overlays three series plots of sales by quarter on a discrete x-axis. The DISCRETEOFFSET option is set for each curve to create a horizontal separation between the curves.

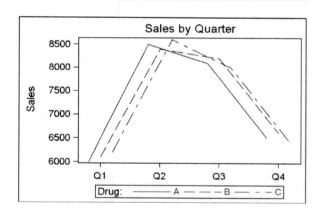

```
title 'Sales by Quarter';
proc sgplot data=sgbook.seriesDiscrete;
   series x=qtr y=salesa / discreteoffset=-0.2;
   series x=qtr y=salesb / discreteoffset= 0.0;
   series x=qtr y=salesc / discreteoffset=+0.2;
   xaxis grid display=(nolabel) offsetmin=0.1 offsetmax=0.1;
   yaxis grid label='Sales';
   keylegend / title='Drug: ';
run;
```

Figure 4.6.9: Series Plot with Group Overlay

This is a graph with a grouped series plot overlaid with a scatter plot that has error bars on a discrete x-axis. The groups are overlaid on the same midpoint values along the x-axis.

This can obscure the data. See alternative arrangement in Figure 4.6.10.

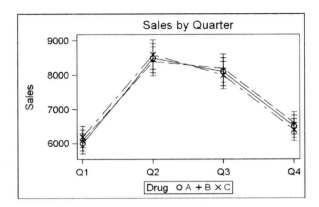

```
title 'Sales by Quarter';
proc sgplot data=sgbook.seriesDiscreteGroup;
   scatter x=qtr y=sales / yerrorupper=high yerrorlower=low
           group=drug;
   series x=qtr y=sales / group=drug;
   xaxis grid display=(nolabel);
   yaxis grid;
   run;
```

Figure 4.6.10: Series Plot with Group Cluster

This graph is the same as in Figure 4.6.9, with GROUPDISPLAY=Cluster. In this case, both the scatter and series plots use the same settings.

This causes the groups to be displayed with an offset along the x-axis, thus making the data for each drug more readable. CLUSTERWIDTH determines the offsets.

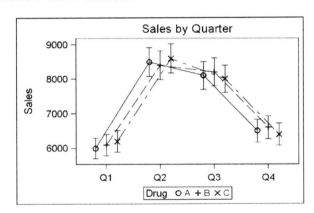

```
title 'Sales by Quarter';
proc sgplot data=sgbook.seriesDiscreteGroup;
   scatter x=qtr y=sales / group=drug yerrorupper=high yerrorlower=low
           groupdisplay=cluster clusterwidth=0.6;
   series x=qtr y=sales / group=drug lineattrs=(thickness=1)
           groupdisplay=cluster clusterwidth=0.6;
   xaxis grid display=(nolabel);
   yaxis grid;
   run;
```

4.7 Step Plot

The step plot is similar to the series plot with a step join for each value. The syntax is:

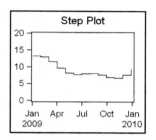

```
step x=column y=column </ options>;
```

A step plot draws a horizontal line through the Y value, with a step change at the next observation.

Required Data Roles:

X	=column	Position along the horizontal axis
Y	=num-column	Position along the vertical axis

Optional Data Roles:

DATALABEL	=column	Label String for each observation
GROUP	=column	Classification for multiple series
URL	=column	URL for each observation
YERRORLOWER	=num-column	Lower value for Y Error Bar
YERRORUPPER	=num-column	Upper value for Y Error Bar

Appearance Options:

ERRORBARATTRS	=line-attrs	Appearance attributes for error bars
LINEATTRS	=line-attrs	Appearance attributes for lines
MARKERATTRS	=marker-attrs	Appearance attributes for markers

Other Options:

CURVELABEL	=string	Curve label for non-grouped case
CURVELABELLOCATION	=keyword	Curve label location (INSIDE | OUTSIDE)
CURVELABELPOSITION	=keyword	Curve label position (TOP, BOTTOM, etc...)
JUSTIFY	=value	Location of data point relative to step

Boolean Options:

BREAK	Break the series plot at missing values
CURVELABEL	Display curve labels from group role
DATALABEL	Display Y value as data label
MARKERS	Display the markers

Standard common options are supported
ATTRID, CLUSTERWIDTH, DISCRETEOFFSET, GROUPDISPLAY, GROUPORDER, LEGENDLABEL, NAME, NOMISSINGGROUP, TRANSPARENCY, URL, X2AXIS, Y2AXIS

Related Style Elements: The GraphDataDefault style element is used to draw the plot elements. When a GROUP variable is present, or when multiple plot statements are overlaid, the GraphData1-GraphData12 elements are used, one for each group or plot statement.

Figure 4.7.1: Basic Step Plot

This is a basic step plot of a response over time for Drug B.

An INSET statement is used to indicate the data is for one of the drugs (B) in the data.

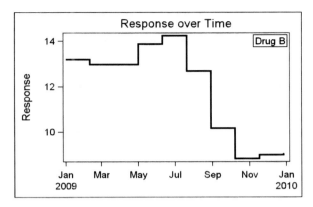

```
title 'Response over Time';
proc sgplot data=sgbook.stepGroup;
   where drug='Drug B';
   step x=date y=val / lineattrs=(thickness=2);
   inset "Drug B" / position=topright border;
   xaxis display=(nolabel);
   yaxis  grid label='Response';
   run;
```

Figure 4.7.2: Grouped Step Plot

This is a grouped step plot of a measure over time for three drugs.

The legend is automatically created for this group case. The KEYLEGEND statement is used to position it inside the data area in a stack.

YAXIS MIN is set to zero.

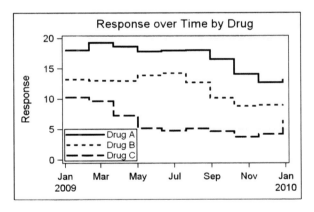

```
title 'Response over Time by Drug';
proc sgplot data=sgbook.stepGroup;
   step x=date y=val / group=drug lineattrs=(thickness=2);
   keylegend / location=inside position=bottomleft  across=1;
   xaxis display=(nolabel);
   yaxis min=0 max=20 integer grid label='Response';
   run;
```

Figure 4.7.3: Grouped Step Plot with Markers

This is a grouped step plot of efficacy over time for three drugs. The LINEATTRS option is used to set the lines patterns to solid.

The markers are displayed. The default justification is left, where the data point is at the left end of the horizontal step. The line first goes horizontally to the right from the data point, then up.

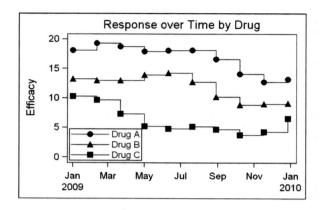

```
title 'Response over Time by Drug';
proc sgplot data=sgbook.stepGroup;
   step x=date y=val / group=drug markers
        lineattrs=(pattern=solid thickness=1);
   keylegend / location=inside position=bottomleft across=1;
   xaxis display=(nolabel);
   yaxis min=0 integer grid label='Efficacy';
run;
```

Figure 4.7.4: Grouped Step Plot with Center Justified Markers

This is a grouped step plot of efficacy over time for three drugs.

The markers are displayed. The JUSTIFY option is set to center. Now the data point is at the center of the horizontal step.

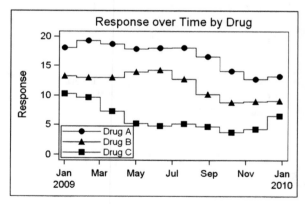

```
title 'Response over Time by Drug';
proc sgplot data=sgbook.stepGroup;
   step x=date y=val / group=drug markers justify=center
        lineattrs=(pattern=solid thickness=1) ;
   keylegend / location=inside position=bottomleft across=1;
   xaxis display=(nolabel);
   yaxis min=0 integer grid label='Efficacy';
   run;
```

Figure 4.7.5: Grouped Step Plot with Upper Limit

This is a step plot of response over time by drug with upper limit values. Note the step plot itself supports options for limits, unlike the series plot, which does not.

The markers are displayed center justified with reduced size. The error bar pattern is the same as the pattern for the line.

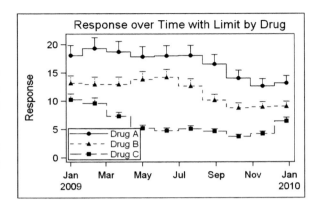

```
title 'Response over Time with Limit by Drug';
proc sgplot data=sgbook.stepGroup;
   step x=date y=val / yerrorupper=upper group=drug  markers
        justify=center markerattrs=(size=5);
   keylegend / location=inside position=bottomleft  across=1;
   xaxis display=(nolabel);
   yaxis min=0 integer grid label='Response';
run;
```

Figure 4.7.6: Grouped Step Plot with Both Limits and Error Bar Attributes

This is a step plot of response over time by drug with both upper and lower limit values. Note the step plot itself supports options for limits, unlike the series plot, which does not.

The markers are displayed center justified with a reduced size. The line pattern is set to solid. The color for the error bars is also set.

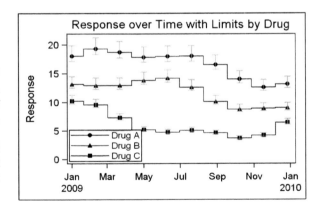

```
title 'Response over Time with Limits by Drug';
proc sgplot data=sgbook.stepGroup;
   step x=date y=val / yerrorupper=upper yerrorlower=lower group=drug
        markers justify=center lineattrs=(pattern=solid)
        markerattrs=(size=5) errorbarattrs=(color=grey);
   keylegend / location=inside position=bottomleft  across=1;
   xaxis display=(nolabel);
   yaxis min=0 integer grid label='Response';
run;
```

Figure 4.7.7: Overlay Step Plot with Curve Labels

Here is an overlay of two step plots with curve labels. CURVELABEL option is used to label the curves. This provides a direct association between name and the curve.

The procedure detects the presence of the CURVELABEL option and suppresses the automatic legend.

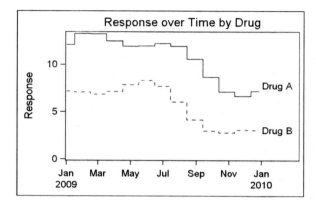

```
title 'Response over Time by Drug';
proc sgplot data=sgbook.step;
  step x=date y=A / justify=center curvelabel='Drug A';
  step x=date y=B / justify=center curvelabel='Drug B';
  xaxis display=(nolabel);
  yaxis min=0 integer grid label='Response';
run;
```

Figure 4.7.8 Overlay Step Plot with Break

Here is the same graph as in Figure 4.7.7, with a break in the data. For Drug A, the BREAK option is not used. For Drug B, the BREAK option is used.

The missing data is between 10Jun and 08Oct, as can be seen by the two reference lines. Drug A plot has one step between the two values. Drug B plot has a break between these values.

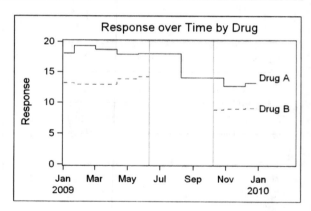

```
title 'Response over Time by Drug';
proc sgplot data=sgbook.stepBreak;
  step x=date y=A / justify=center curvelabel='Drug A';
  step x=date y=B / justify=center curvelabel='Drug B' break;
  refline '10jun2009'd '08oct2009'd / axis=x;
  xaxis display=(nolabel);
  yaxis min=0 integer grid label='Response';
run;
```

4.8 Band Plot

The band plot draws a band between two response values. The syntax can be one of the following:

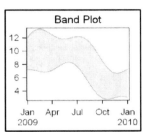

```
band x=column lower=column|value upper= column|value
    </ options>;
band y=column lower=column|value upper= column|value
    </ options>;
```

For each value of the X (or Y) variable, a band is drawn between the lower and upper response values. Response values can come from a column in the data or can be a constant value.

Required Data Roles:

X or Y	=column	Position along horizontal or vertical axis
LOWER	=num-column | value	Lower response
UPPER	=num-column | value	Upper response

Optional Data Roles:

GROUP	=column	Classification for multiple series

Appearance Options:

CURVELABELATTRS	=text-attrs	Appearance of curve label
FILLATTRS	=fill-attrs	Appearance of fill between values
LINEATTRS	=line-attrs	Appearance of the lines

Other Options:

CURVELABELLOC	=keyword	Location of label – INSIDE | OUTSIDE
CURVELABELPOS	=keyword	Position of label - START | END
CURVELABELLOWER	=string	Name of lower band curve
CURVELABELUPPER	=string	Name of upper band curve
MODELNAME	=plotname	Name of the plot for interpolation info.
NAME	=string	Name for this statement
TRANSPARENCY	=value	Transparency for the line and markers
TYPE	=value	Series or Step

Boolean Options:

FILL | NOFILL	Display fill between lower and upper values
OUTLINE | NOOUTLINE	Display outline at lower and upper values
NOEXTEND	

Standard common options are supported
ATTRID, LEGENDLABEL, NAME, NOMISSINGGROUP, TRANSPARENCY, X2AXIS, Y2AXIS

Related Style Elements: The GraphDataDefault style element is used to draw the plot elements. When a GROUP variable is present, or when multiple plot statements are overlaid, the GraphData1-GraphData12 elements are used, one for each group or plot statement.

The band may be filled, with or without an outline. The GROUP role is used to create multiple bands, one for each unique group value. When used with a series or step plot to display a confidence or prediction band, the MODELNAME option can be used to create an association with the fit plot.

Note: The band plot statement is optimal for the display of confidence and prediction bands, where the X (or Y) values are sorted. The band is drawn in data order, and if the data are not sorted by X (or Y), then the results may be unpredictable.

Figure 4.8.1: Basic Band Plot

This is a basic band plot along the x-axis, with upper and lower limits in the y-axis direction.

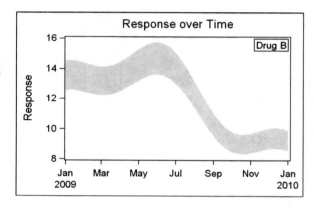

```
title 'Response over Time';
proc sgplot data=sgbook.bandGroup;
  where drug='Drug B';
  band x=date upper=upper lower=lower;
  inset "Drug B" / position=topright border;
  xaxis display=(nolabel);
  yaxis  grid label='Response';
run;
```

Figure 4.8.2: Grouped Band Plot

This is a band plot of upper and lower Y response by X and drug. The legend is automatically created.

The x-axis label is suppressed.

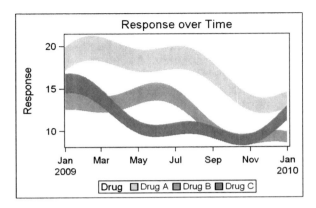

```
title 'Response over Time';
proc sgplot data=sgbook.bandGroup;
  band x=date upper=upper lower=lower / group=drug fill;
  xaxis display=(nolabel);
  yaxis grid label='Response';
  run;
```

Figure 4.8.3: Grouped Band Plot with Transparency

This is a band plot of upper and lower Y response by X and drug. The legend is automatically created.

Use of transparency allows the viewing of the overlapped regions of the bands.

The x-axis label is suppressed.

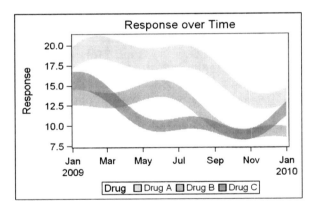

```
title 'Response over Time';
proc sgplot data=sgbook.bandGroup;
  band x=date upper=upper lower=lower / group=drug transparency=0.3;
  xaxis display=(nolabel);
  yaxis grid label='Response';
  run;
```

Figure 4.8.4: Overlay Band Plots

This is an overlay of two band plots with independent columns for upper and lower responses by date for each drug. The individual bands statements can be assigned different visual attributes.

The legend is automatically created. The legend entries are derived from the LEGENDLABEL settings for each plot.

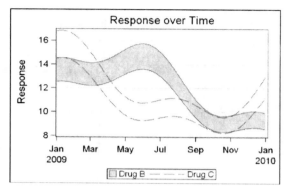

```
title 'Response over Time';
proc sgplot data=sgbook.bandBreak;
  band x=date upper=upperB lower=lowerB / transparency=0.3
        fill outline legendlabel='Drug B';
  band x=date upper=upperC lower=lowerC / transparency=0.3
        outline legendlabel='Drug C';
  xaxis display=(nolabel);
  yaxis grid label='Response';
  run;
```

Figure 4.8.5: Overlay Band Plots with Constant Lower Limit

This is an overlay of two band plots with independent columns for upper and lower responses by date for each drug. The first band has LOWER=0 (constant).

A "fill under line" effect can be created using a band in this way. LEGENDLABEL options are used to name the bands in the legend.

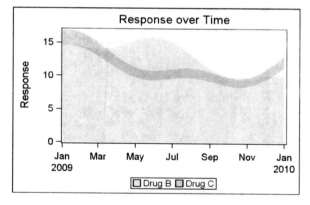

```
title 'Response over Time';
proc sgplot data=sgbook.bandBreak;
  band x=date upper=upperB lower=0 / transparency=0.5
        legendlabel='Drug B';
  band x=date upper=upperC lower=lowerC / transparency=0.5
        legendlabel='Drug C';
  xaxis display=(nolabel);
  yaxis grid label='Response';
  run;
```

Figure 4.8.6: Overlay Band Plots with Curve Labels

This is a graph with two overlaid band plots. The CURVELABEL options are used to display the upper and lower labels for one of the band plots.

CURVELABELPOS option can be set to start or end. Since 'end' is the default setting, it need not be specified in this program.

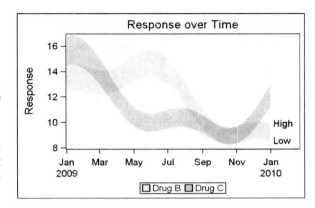

```
title 'Response over Time';
proc sgplot data=sgbook.bandBreak;
  band x=date upper=upperB lower=lowerB / transparency=0.5
    curvelabelupper='High' curvelabellower='Low' curvelabelpos=end
    legendlabel='Drug B';
  band x=date upper=upperC lower=lowerC / transparency=0.5
    legendlabel='Drug C';
  xaxis display=(nolabel);
  yaxis grid label='Response';   run;
```

Figure 4.8.7: Overlay Step and Band Plots

This graph shows a step plot of the response value along with a band plot of the limits.

MODELNAME="Value" is assigned on the band plot, where "Value" is the name of the step plot.

The Band plot derives its join characteristics, such as the step shape and the justification from the step plot.

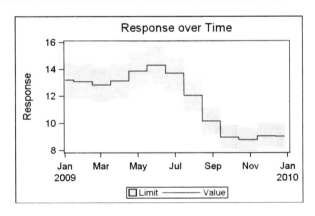

```
title 'Response over Time';
proc sgplot data=sgbook.stepBand;
  band x=date upper=upperB lower=lowerB / transparency=0.5
       modelname='Value' legendlabel='Limit';
  step x=date y=B / justify=center name='Value' legendlabel='Value';
  xaxis display=(nolabel);
  yaxis grid label='Response';
  run;
```

4.9 Needle Plot

The needle plot draws a vertical needle for each observation from the (x,y) coordinate to the x-axis. The syntax is:

```
needle x=column y=num-column </ options>;
```

A vertical line (needle) is drawn from each observation to the baseline. For Interval x-axis, the GROUP role affects only the display attributes of the needle. Cluster groups are supported for discrete axis.

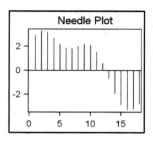

Required Data Roles:

X	=column	Position along the horizontal axis
Y	=num-column	Position along the vertical axis

Optional Data Roles:

DATALABEL	=column	Label String for each observation
GROUP	=column	Classification for multiple series
URL	=column	URL for each observation

Appearance Options:

DATALABELATTRS	=text-attrs	Appearance attributes for labels
LINEATTRS	=line-attrs	Appearance attributes for the needles
MARKERATTRS	=marker-attrs	Appearance attributes for markers

Other Options:

BASELINE	=number	Y-intercept for the baseline

Boolean Options:

DATALABEL	Display Y value as data label
MARKERS	Display the markers

Standard common options are supported

ATTRID, DISCRETEOFFSET, GROUPDISPLAY, GROUPORDER, LEGENDLABEL, NAME, NOMISSINGGROUP, TRANSPARENCY, URL, X2AXIS, Y2AXIS

Related Style Elements: The GraphDataDefault style element is used to draw the plot elements. When a GROUP variable is present, or when multiple plot statements are overlaid, the GraphData1-GraphData12 elements are used, one for each group or plot statement.

Figure 4.9.1: Basic Needle Plot

This is a basic needle plot of trading volume by month.

LINEATTRS option is used to set the needle thickness to 3px.

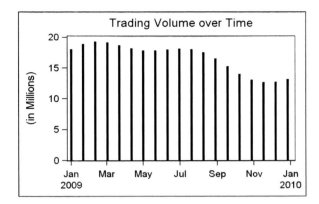

```
title 'Trading Volume over Time';
proc sgplot data=sgbook.needleTrade;
   needle x=date y=val / lineattrs=(thickness=3);
   xaxis display=(nolabel);
   yaxis  grid label='(in Millions)';
   run;
```

Figure 4.9.2: Grouped Needle Plot

This is a grouped needle plot of trading volume by month. The trade is either a "Buy" or a "Sell". This is used as the GROUP role.

An OFFSETMAX=0.1 is set for y-axis to make room for legend that has been placed inside.

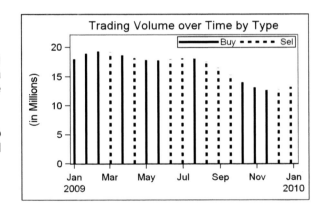

```
title 'Trading Volume over Time by Type';
proc sgplot data=sgbook.needleTrade;
   needle x=date y=val / lineattrs=(thickness=3) group=trade;
   xaxis display=(nolabel);
   yaxis  grid label='(in Millions)' offsetmax=0.1;
   keylegend / location=inside position=topright;
   run;
```

Figure 4.9.3: Needle Plot with a Baseline

This is a needle plot of trades over time.
BASELINE=15 is set, so needles are drawn
to this baseline.

Markers and data labels are displayed.
Data labels are automatically moved to
avoid collision with other labels

Needle thickness is set to 3px.

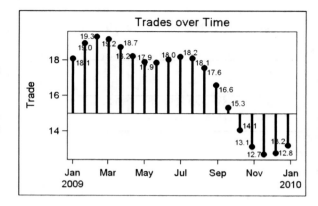

```
title 'Trades over Time';
proc sgplot data=sgbook.needleTrade;
   needle x=date y=val / lineattrs=(thickness=3) baseline=15
          datalabel=val markers;
   xaxis display=(nolabel);
   yaxis grid label='Trade';
   run;
```

Figure 4.9.4: Needle Plot with Discrete Groups

This is a needle plot of trades by month,
grouped by trade type (Buy or Sell).

In this case, the x-axis is discrete, so the
groups can be displayed as clusters. The
CLUSTERWIDTH option is used to reduce
the cluster width to get tighter clustering.

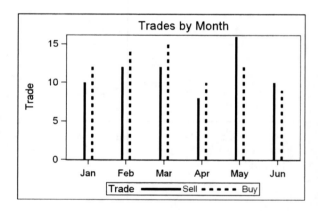

```
title 'Trades by Month';
proc sgplot data=sgbook.needleDiscrete;
   needle x=month y=val / group=trade lineattrs=(thickness=3)
          groupdisplay=cluster clusterwidth=0.4;
   xaxis display=(nolabel);
   yaxis grid label='Trade';
   run;
```

4.10 Vector Plot

The Vector plot draws directional vectors. The syntax is:

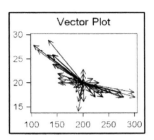

```
vector x=column y=column < / options>;
```

A vector is drawn to each observation from the origin (0,0) by default. The origin may be set by using the XORIGIN and YORIGIN roles that can have data columns or constants.

Required Data Roles:
```
    X                   =num-column     Position of arrow head along the x-axis
    Y                   =num-column     Position of arrow head along the y-axis
```

Optional Data Roles:
```
    DATALABEL           =column                 Label string for each vector
    GROUP               =column                 Classification variable
    XORIGIN             =num-column | value     Position of arrow tail along the x-axis
    YORIGIN             =num-column | value     Position of arrow tail along the y-axis
```

Appearance Options:
```
    DATALABELATTRS      =text-attrs     Appearance attributes for data labels
    LINEATTRS           =line-attrs     Appearance attributes for lines
```

Other Options:
```
    ARROWHEADDIRECTION  =keyword    OUT | IN | BOTH
    ARROWHEADSHAPE      =keyword    OPEN | CLOSED | FILLED | BARBED
```

Boolean Options:
```
    DATALABEL                       Display curve labels from group role
    NOARROWHEADS                    Suppress display of arrow heads
```

Standard common options are supported
    ATTRID, LEGENDLABEL, NAME, NOMISSINGGROUP, TRANSPARENCY, X2AXIS, Y2AXIS

Related Style Elements: The GraphDataDefault style element is used to draw the plot elements. When a GROUP variable is present, or when multiple plot statements are overlaid, the GraphData1-GraphData12 elements are used, one for each group or plot statement.

Figure 4.10.1: Basic Vector Plot

This is the basic vector plot with default origin of (0, 0).

A vector is drawn from the origin to the (x,y) location for each observation.

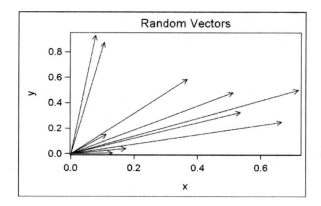

```
title 'Random Vectors';
proc sgplot data=sgbook.vector;
   vector x=x y=y;
   run;
```

Figure 4.10.2: Vector Plot

This is a vector plot where all the vectors have a user defined common constant origin (0.4, 0.4) specified by the XORIGIN and YORIGIN roles.

The LINEATTRS option is used to set the line thickness to 2. Note the arrowhead sizes are proportional to the line thickness.

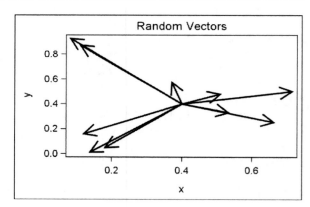

```
title 'Random Vectors';
proc sgplot data=sgbook.vector;
   vector x=x y=y / xorigin=0.4 yorigin=0.4 lineattrs=(thickness=2);
   run;
```

Figure 4.10.3: Vertical Vector Plot

This stock plot is created by using vectors to draw the monthly closing stock price over time

ARROWHEADSHAPE option is used to get barbed arrowheads.

SAS 9.3 supports a new HIGHLOW plot, which provides another way to create such a graph.

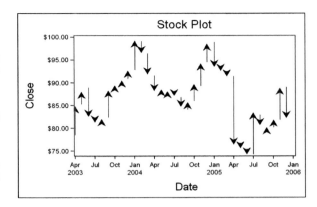

```
title 'Stock Plot';
proc sgplot data=sgbook.Stock;
   vector x=date y=close / xorigin=date yorigin=open
          arrowheadshape=barbed;
   xaxis grid;
   yaxis grid;
   run;
```

Figure 4.10.4: Grouped Vector Plot

This graph uses a grouped vector plot where each vector has its own origin. Increasing and decreasing trend have different groups. Groups are represented with different visual attributes.

NOARROWHEADS option is used.

SAS 9.3 supports a new HighLow plot, which provides another way to create such a graph.

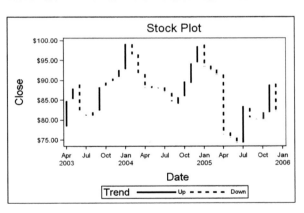

```
title 'Stock Plot';
proc sgplot data=sgbook.Stock;
   vector x=date y=close / xorigin=date yorigin=open group=trend
          noarrowheads lineattrs=(thickness=2);
   xaxis grid;
   yaxis grid;
   run;
```

4.11 VBarParm and HBarParm Plots (9.3)

These pre-summarized bar chart statements have the following syntax:

```
vbarparm column < / response=column> <options>;
hbarparm column < / response=column> <options>;
```

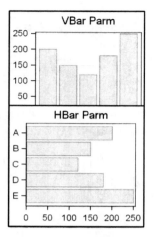

The VBARPARM and HBARPARM statements create specialized bar charts in which the response and limit values are displayed as provided by the user.

Related Style Elements: The GraphDataDefault style element is used to draw the plot elements. When a GROUP variable is present, or when multiple plot statements are overlaid, the GraphData1-GraphData12 elements are used, one for each group or plot statement.

Required Data Roles:
Category	=column	Position along the category axis
Response	=num-column	Response variable

Optional Data Roles:
DATALABEL	=column	Label String for each observation
GROUP	=column	Classification for multiple series
URL	=column	URL for each observation
LIMITLOWER	=column	Lower limit value
LIMITUPPER	=column	Upper limit value

Appearance Options:
FILLATTRS	=fill-attrs	Appearance of bar interior
LIMITATTRS	=line-attrs	Appearance of limit bars

Other Options:
BARWIDTH	= value	Width of bar as fraction of tick spacing
DATASKIN	= skin-value	One of the predefined skin types

Boolean Options:
DATALABEL	Display default bar data labels
FILL | NOFILL	Display filled / unfilled bars
OUTLINE | NOOUTLINE	Display bar outlines or not.
MISSING	Accept missing as category value

Standard common options are supported
ATTRID, CLUSTERWIDTH, DISCRETEOFFSET, GROUPDISPLAY, GROUPORDER, LEGENDLABEL, NAME, NOMISSINGGROUP, TRANSPARENCY, URL

Figure 4.11.1: Basic Parametric Bar Chart

This is a basic bar chart of mean MPG (City) for all cars by origin. The data has been summarized using the MEANS procedure to create the data set needed for using this plot statement.

Since the data was summarized by origin and type, we have selected _TYPE_=2 for summary by origin.

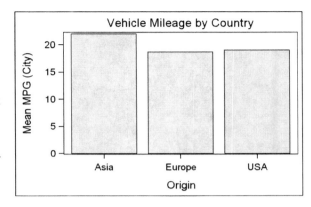

```
title 'Vehicle Mileage by Country';
proc sgplot data=sgbook.carMeans;
  where _type_=2;
  vbarparm category=origin response=mean_mpg_city;
  run;
```

Figure 4.11.2: Parametric Bar Chart with Limits and Data Labels

This is a vertical bar chart of mean MPG (City) for all cars with limits and labels by Type.

Because the limits are displayed, the data label values are displayed below the bars. The labels can be moved using the DATALABELPOS option.

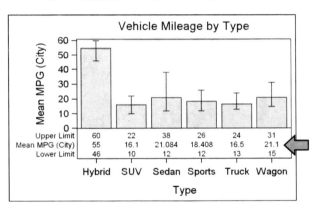

```
title 'Vehicle Mileage by Type';
proc sgplot data=sgbook.carMeans;
  where _type_=1;
  vbarparm category=type response=mean_mpg_City / datalabel
         limitupper=max_mpg_city limitlower=min_mpg_city
           datalabelpos=bottom;
  yaxis grid;
  run;
```

Figure 4.11.3: Grouped Parametric Bar Chart

This is a grouped bar chart of mileage by origin and type. GROUPDISPLAY is set to Cluster for the side-by-side display of the group values.

This bar chart uses DATASKIN=sheen.

YAXIS OFFSETMAX is set to create some room for the legend inside the data area.

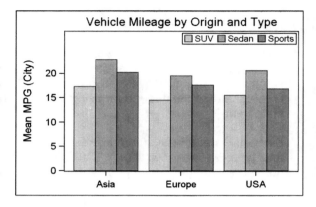

```
title 'Vehicle Mileage by Origin and Type';
proc sgplot data=sgbook.carMeans;
  where _type_=3 and (type = 'Sedan' or type='Sports' or type='SUV');
  vbarparm category=origin response=mean_mpg_City / group=type
           transparency=0.2;
  keylegend / location=inside position=topright;
  xaxis display=(nolabel); yaxis grid offsetmax=0.2;
run;
```

Figure 4.11.4: Grouped Parametric Bar Chart with Data Labels

This grouped bar chart by type is the same as in Figure 4.11.3. The DATALABEL option is used to display the bar statistics.

A format is set to reduce the precision on the data labels. DATASKIN=sheen.

YAXIS OFFSETMAX=0.2 is set to make room for the legend.

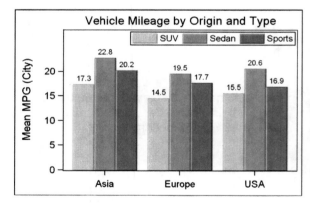

```
title 'Vehicle Mileage by Origin and Type';
proc sgplot data=sgbook.carMeans;
  where _type_=3 and (type = 'Sedan' or type='Sports' or type='SUV');
  format mean_mpg_city 5.1;
  vbarparm category=origin response=mean_mpg_City / group=type
           dataskin=sheen datalabel;
  keylegend / location=inside position=topright;
  xaxis display=(nolabel);
  yaxis grid offsetmax=0.2;   run;
```

Figure 4.11.5: Overlay Parametric Bar Charts

This graph uses overlay of two VBARPARM statements, one each for the City and Highway mileage.

For the common categories, the bars of the second statement are overlaid directly over the first, so it is better to set the BARWIDTH of the second bar to a smaller value.

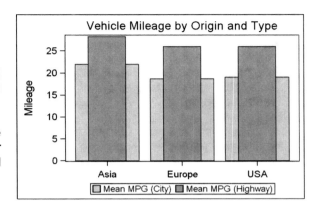

```
title 'Vehicle Mileage by Origin';
proc sgplot data=sgbook.carMeans;
  where _type_=2;
  vbarparm category=origin response=mean_mpg_City;
  vbarparm category=origin response=mean_mpg_Highway / barwidth=0.5;
  xaxis display=(nolabel);
  yaxis grid label='Mileage';   run;
```

Figure 4.11.6: Overlay Parametric Bar Charts with Skins and Offsets

Overlay VBARPARM statements, one each for City and Highway mileage.

Each set of bars is offset from the midpoint by setting the appropriate value for the DISCRETEOFFSET option. Bar widths are reduced appropriately. The values are a fraction of the category spacing.

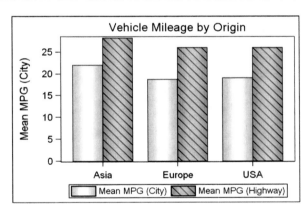

```
ods listing style=journal3;
title 'Vehicle Mileage by Origin';
proc sgplot data=sgbook.carMeans;
  where _type_=2;
  vbarparm category=origin response=mean_mpg_City /
          dataskin=pressed discreteoffset=-0.2 barwidth=0.4;
  vbarparm category=origin response=mean_mpg_Highway /
          dataskin=pressed discreteoffset=0.2 barwidth=0.4;
  xaxis display=(nolabel);
  yaxis grid;
  run;
```

Figure 4.11.7: Horizontal Parametric Bar Chart

This graph uses an HBARPARM plot to display the mean frequency of adverse events for a drug.

Horizontal bars are particularly useful when the category names are long.

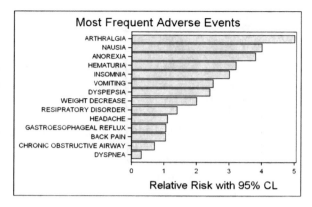

```
title 'Most Frequent Adverse Events';
proc sgplot data=sgbook.AdverseEvents;
  hbarparm category=ae response=mean;
  yaxis display=(nolabel);
  xaxis grid label='Relative Risk with 95% CL';
run;
```

Figure 4.11.8: Horizontal Parametric Bar Chart with Limits

This graph uses an HBARPARM plot to display the mean frequency of adverse events for a drug with low and high confidence limits.

We use the LIMITATTRS option to set the visual attributes of the limit bars.

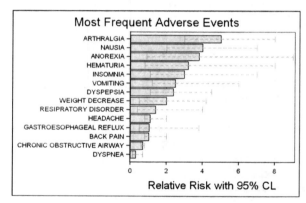

```
title 'Most Frequent Adverse Events';
proc sgplot data=sgbook.AdverseEvents;
  hbarparm category=ae response=mean / limitlower=low limitupper=high
        limitattrs=(color=grey pattern=dash);
  yaxis display=(nolabel);
  xaxis grid label='Relative Risk with 95% CL';
run;
```

4.12 Bubble Plot (9.3)

The BUBBLE statement has the following syntax:

```
bubble x=column y=column size=num-column </ options>;
```

Bubble plot draws circular bubbles sized by the Size role at the X and Y locations in the plot. Radius of the bubbles is scaled linearly by the SIZE variable.

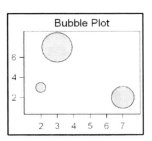

Required Data Roles:
X	*column*	Position along the horizontal axis
Y	*column*	Position along the vertical axis
SIZE	*num-column*	Size of the bubble

Optional Data Roles:
DATALABEL	*=column*	Label String for each observation
GROUP	*=column*	Classification for bubble color
URL	*=column*	URL for each observation

Appearance Options:
DATALABELATTRS	*=text-attrs*	Appearance attributes of the labels
FILLATTRS	*=fill-attrs*	Appearance attributes of bubble interior
LINEATTRS	*=line-attrs*	Appearance attributes for bubble outline

Other Options:
BRADIUSMAX	*=value*	Radius in pixels of the largest bubble
BRADIUSMIN	*=value*	Radius in pixels of the smallest bubble

Boolean Options:
FILL	NOFILL	Display filled bubbles or not
OUTLINE	NOOUTLINE	Display bubble outline or not

Standard common options are supported
ATTRID, LEGENDLABEL, NAME, NOMISSINGGROUP, TRANSPARENCY, URL, X2AXIS, Y2AXIS

Related Style Elements: The GraphDataDefault style element is used to draw the plot elements. When a GROUP variable is present, or when multiple plot statements are overlaid, the GraphData1-GraphData12 elements are used, one for each group or plot statement.

Figure 4.12.1: Basic Bubble Plot

This is a basic bubble plot of Age by Height and Weight. This graph is for female students in the sashelp.class data set.

An INSET statement is used to identify the subset of the data.

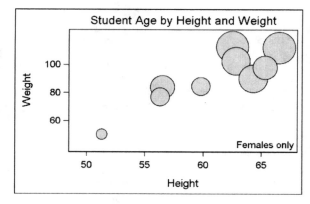

```
title 'Student Age by Height and Weight';
proc sgplot data=sashelp.class;
  where sex='F';
  bubble x=height y=weight size=age;
  inset 'Females only' / position=bottomright;
  xaxis grid;   yaxis grid;
  run;
```

Figure 4.12.2: Grouped Bubble Plot

This is a grouped bubble plot of horsepower by City and Highway Mileage for car statistics summarized by Type, excluding Hybrids.

The KEYLEGEND statement is used to position the legend in the bottom right corner of the data area.

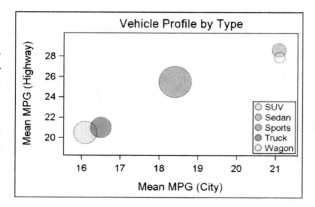

```
title 'Vehicle Profile by Type';
proc sgplot data=sgbook.carMeans;
  where _type_ = 1 and type <> 'Hybrid';
  bubble x=mean_mpg_city y=mean_mpg_highway size=mean_horsepower /
     group=type transparency=0.5;
  keylegend / location=inside position=bottomright across=1;
  xaxis grid;   yaxis grid;
  run;
```

Figure 4.12.3: Grouped Bubble Plot

This graph is the same as in Figure 4.12.2 with each bubble labeled by the GROUP variable.

The NOAUTOLEGEND option is used to remove the automatic legend, since direct labeling is preferred to having a separate legend.

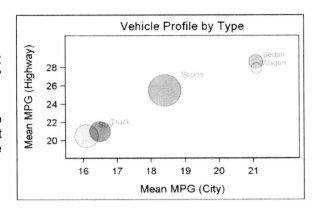

```
title 'Vehicle Profile by Type';
proc sgplot data=sgbook.carMeans noautolegend;
  where _type_ = 1 and type <> 'Hybrid';
  bubble x=mean_mpg_city y=mean_mpg_highway size=mean_horsepower /
         group=type transparency=0.5 datalabel=type;
  xaxis grid;  yaxis grid;
run;
```

Figure 4.12.4: Bubble Plot with Negative Response Data

This Bubble plot represents negative size values using a dash pattern. This is achieved by using the absolute values for the size role. The GROUP variable is set based on the sign of the data and used with the group role.

Bubble radius min and max are set to map the data proportionately. A scatter plot with MARKERCHAR is used for the labels.

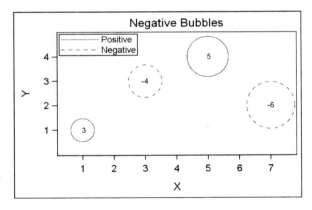

```
title 'Negative Bubbles';
proc sgplot data=sgbook.bubble2;
  bubble x=x y=y size=AbsValue / group=group transparency=0.2
         bradiusmin=15  bradiusmax=30 nofill name='bubble';
  scatter x=x y=y / markerchar=value;
  xaxis grid integer;   yaxis grid integer;
  keylegend 'bubble' / location=inside position=topleft across=1;
run;
```

4.13 HighLow Plot (9.3)

The HIGHLOW plot statement has the following syntax:

```
highlow x=column low=num-column high=num-column </opts>;
highlow y=column low=num-column high=num-column </opts>;
```

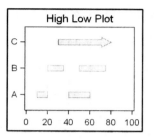

This is a versatile plot type that displays a vertical or horizontal high-low plot, using either a "bar" or "line" visual. The line version is useful to create traditional stock plots. The bar version is useful for event timelines.

Required Data Roles:

X or Y	*column*	Position along the vertical or horizontal axis
HIGH	*num-column*	End point of the bar or line segment
LOW	*num-column*	Start point of the bar or line segment

Optional Data Roles:

CLOSE	=*num-column*	Close value for bar or line
GROUP	=*column*	Classification for multiple series
HIGHCAP | LOWCAP	=*column* | *keyword*	Caps (NONE, OPENARROW, etc.)
HIGHLABEL | LOWLABEL	=*column* | *string*	Label for the upper or lower end
OPEN	=*num-column*	Open value for bar or line
URL	=*column*	URL for each observation

Appearance Options:

FILLATTRS	=*fill-attrs*	Appearance of bar interior
DATALABELATTRS	=*text-attrs*	Appearance of bar labels
LINEATTRS	=*line-attrs*	Appearance of line or outline

Other Options:

BARWIDTH	=*value*	Width of bar as fraction of tick spacing
GROUPDISPLAY	=*value*	Cluster or Overlay
GROUPORDER	=*keyword*	DATA | ASCENDING | DESCENDING
INTERVALBARWIDTH	=*value*	Width of bar in pixels
TYPE	=*keyword*	BAR | LINE

Boolean Options:

FILL | NOFILL	Display filled bars or not for Type=Bar
OUTLINE | NOOUTLINE	Display outline or not for Type=Bar

Standard common options are supported
ATTRID, CLUSTERWIDTH, DISCRETEOFFSET, GROUPDISPLAY, GROUPORDER, LEGENDLABEL, NAME, NOMISSINGGROUP, TRANSPARENCY, X2AXIS, Y2AXIS

Related Style Elements: Same as for the VBarParm Plot statement.

Figure 4.13.1: HIGHLOW Line Plot

This graph uses the HighLow plot to create a Stock Plot of prices by month.

The LOW and HIGH roles are used to display the open and close values. The default TYPE is line.

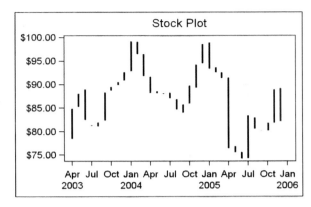

```
title 'Stock Plot';
proc sgplot data=sgbook.stock;
  highlow x=date low=open high=close / lineattrs=(thickness=2);
  xaxis grid display=(nolabel);
  yaxis grid label='Price';
  run;
```

Figure 4.13.2: HighLow Line Plot with Open and Close Values

This graph uses the HighLow plot to create a Stock Plot of prices by month.

The LOW and HIGH roles are used to display the monthly low and high values. The OPEN and CLOSE roles are used to display the open and close values. The default TYPE is line.

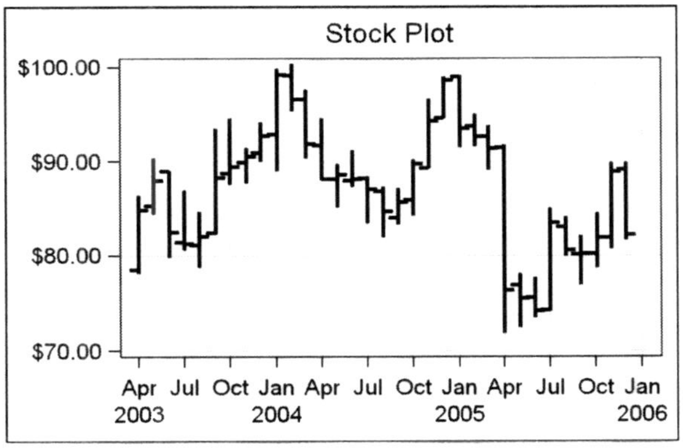

```
title 'Stock Plot';
proc sgplot data=sgbook.stock;
  highlow x=date low=low high=high / open=open close=close
          lineattrs=(thickness=2);
  xaxis grid display=(nolabel);
  yaxis grid display=(nolabel);
  run;
```

Figure 4.13.3: HighLowLine Plot with End Caps

This is a basic Stock Plot showing the open and close stock prices by Month. The trend is shown using the end caps.

Note: if the line segment is not long enough, end caps are not shown.

Lower is the smaller of Open and Close. Higher is the greater of Open and Close.

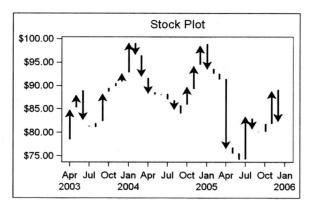

```
title 'Stock Plot';
proc sgplot data=sgbook.stockcap;
  highlow x=date low=lower high=higher / lineattrs=(thickness=2)
          highcap=highcap lowcap=lowcap;
  xaxis grid display=(nolabel);
  yaxis grid display=(nolabel);
  run;
```

Figure 4.13.4: HighLow Bar Chart

This is a HighLow plot of mileage range by type. TYPE=Bar, and BarWidth = 0.5.

LOWLABEL and HIGHLABEL options are used to display the low and high labels from specific columns.

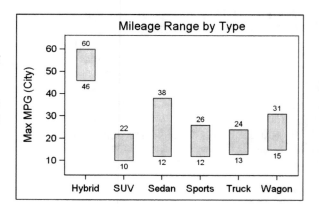

```
title 'Mileage Range by Type';
proc sgplot data=sgbook.carMeans;
  where _type_ = 1;
  highlow x=type low=min_mpg_City high=max_mpg_city / type=bar
          barwidth=0.5 lowlabel=min_mpg_city highlabel=max_mpg_city;
  xaxis grid display=(nolabel);
  yaxis grid;
  run;
```

Figure 4.13.5: HighLow Bar Chart

This is an Adverse Event Timeline for a patient in a drug study. Data comes in CDISC format. Events are plotted by aeseq. The LOWLABEL role is used to display the aedecod names. The y-axis is discrete.

For two of the events the aeendate is missing, so the bar segment is not drawn. This is indicated by the annotated arrows.

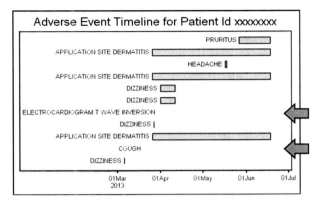

```
title 'Adverse Event Timeline for Patient Id xxxxxxxx';
proc sgplot data=sgbook.AETimeline tmplout='highlow.sas';
  highlow y=aeseq low=aestdate high=aeendate / type=bar barwidth=0.6
          lowlabel=aedecod;
  xaxis grid display=(nolabel);
  yaxis grid type=discrete display=(noticks novalues nolabel);
  run;
```

Figure 4.13.6: HighLow Bar Chart

This graph is same as in Figure 4.13.5.

For the two events with missing aeendate, we have inserted the study end date for them, and marked the obs to display a HighCap, These events are shown with a high cap arrow, indicating events continuing past the study end date.

The bar width includes the cap.

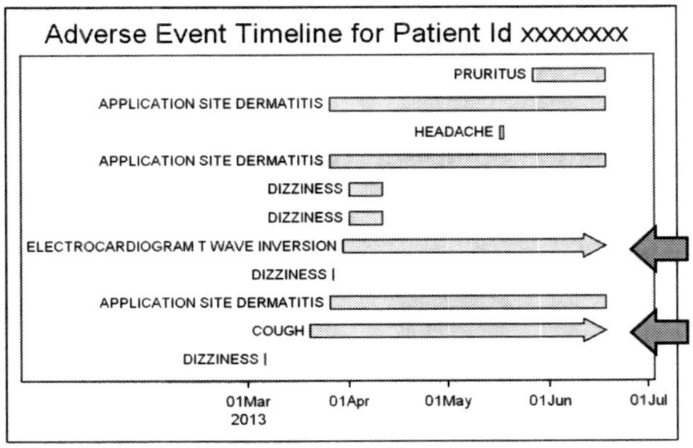

```
title 'Adverse Event Timeline for Patient Id xxxxxxxx';
proc sgplot data=sgbook.AETimelinecap tmplout='highlow.sas';
  highlow y=aeseq low=aestdate high=aeendate / type=bar barwidth=0.8
          lowlabel=aedecod highcap=highcap;
  xaxis grid display=(nolabel);
  yaxis grid display=(noticks novalues nolabel);
  run;
```

4.14 Reference Lines

X and Y reference lines have the following syntax:

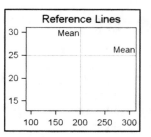

```
refline value <value> </ options>;
```

One or more vertical or horizontal reference lines can be added to a graph using the above syntax. Reflines can be aligned to any one of the four axes X, Y, X2 or Y2. The GraphReference style element is used.

Required Data Roles:

	Value-list	Position along the x-axis or y-axis

Appearance Options:

LINEATTRS	=line-attrs	Appearance attributes for the reference lines

Other Options:

AXIS	= axis	The axis X, X2, Y, Y2 for the values
LABEL	= string	Labels for the reference line(s)
LABELLOC	= string	Location of the label (Inside | Outside)
LABELPOS	= string	Position of the label (Start | End)

Boolean Options:

NOCLIP	Refline is included to determine axis range

Common options supported
DISCRETEOFFSET, LEGENDLABEL, NAME, TRANSPARENCY

Related Style Elements: The GraphReference style element is used to draw these lines.

Figure 4.14.1: Scatter with Reference Line

In this example, we have added a Y reference line with default settings. The associated label is displayed outside the graph, on the right side.

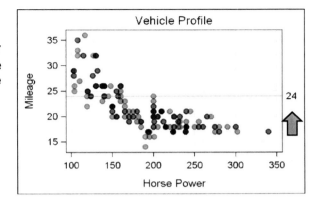

```
title 'Vehicle Profile';
proc sgplot data=sgbook.cars2;
  where type='Sedan' and (origin ='USA' or origin='Asia');
  scatter x=hp y=mpg / transparency=0.5;
  refline 24 / label='24';
  xaxis grid;  yaxis grid;
  run;
```

Figure 4.14.2: Scatter with Reference Line

In this graph the scatter plot response is plotted on the Y2 axis (right side). By default, the reference line label would have been shown on the right side, thus colliding with the Y2 axis.

Here we have set LABELPOS=MIN, to place the label on the left side.

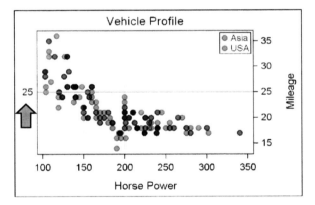

```
title 'Vehicle Profile';
proc sgplot data=sgbook.cars2 ;
  where type='Sedan' and (origin ='USA' or origin='Asia');
  scatter x=hp y=mpg / group=origin transparency=0.5
          markerattrs=(symbol=circlefilled) y2axis;
  keylegend / location=inside position=topright across=1;
  refline 25 / axis=y2 label='25' labelpos=min;
  xaxis grid;  yaxis grid;
  run;
```

Figure 4.14.3: REFLINE on Discrete Axis

This is a bar chart of mean MPG (City) for all cars by Type. We want to place a vertical reference line between 'Sports' and 'Truck' on the x-axis.

We do this by providing the category value, AXIS=X and DISCRETEOFFSET=-0.5. This places the reference line midway between the specified and previous midpoint value.

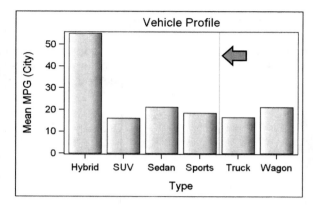

```
title 'Vehicle Profile';
proc sgplot data=sgbook.carMeans ;
  where _type_ = 1;
  vbarparm category=type response=mean_mpg_city / dataskin=sheen;
  refline 'Truck' / axis=x discreteoffset=-0.5;
  yaxis grid;
run;
```

Figure 4.14.4: Multiple Reference Lines

In this example, we have added multiple Y reference lines with inside labels positioned at the lower side of the axis. We have set XAXIS OFFSETMIN to create some room.

We have added an X reference line at x=425. Since this value is beyond the data range, this refline would normally get clipped. To retain it, we have specified the NoClip option.

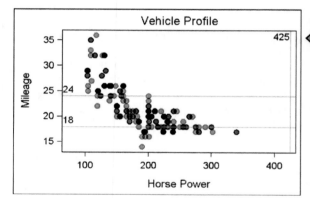

```
title 'Vehicle Profile';
proc sgplot data=sgbook.cars2;
  where type='Sedan' and (origin ='USA' or origin='Asia';
  scatter x=hp y=mpg / transparency=0.5;
  refline 18 24 / label=('18' '24') labelloc=inside labelpos=min;
  refline 425 / axis=x label='425' labelloc=inside noclip;
  xaxis offsetmin=0.1 grid;  yaxis grid;
run;
```

4.15 Parametric Line Plot (9.3)

This parametric line plot statements has the following syntax:

```
lineparm x=column y=column slope=num-column </ opts>;
```

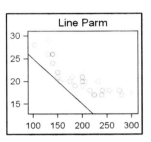

The LINEPARM statement draws an infinitely long line given a point and slope. The plot cannot be used by itself as it is unbounded. Graph data extents are set by other data driven plots in the graph.

Required Data Roles:

X	column	X coordinate of point
Y	column	Y coordinate of point
SLOPE	column	Slope of the line. Missing implies a vertical line.

Optional Data Roles:

CURVELABEL	=column	Label for the line
GROUP	=column	Classification for multiple series
INDEX	=num-column	Index for visual attributes

Appearance Options:

CURVELABELATTRS	=text-attrs	Appearance of the label
CURVELABELLOC	=keyword	OUTSIDE	INSIDE
CURVELABELPOS	=keyword	AUTO	MAX	MIN
LINEATTRS	=line-attrs	Appearance of the line

Other Options:

CLIP	=boolean	Include data for this plot in the axis range
LEGENDLABEL	=string	Label that appears in legend
NAME	=string	Name for this statement

Boolean Options:

NOEXTEND	Line is not drawn to the axes
X2AXIS	Assigns X values to the X2 (top) axis
Y2AXIS	Assigns Y values to the Y2 (right) axis

Standard common options are supported
ATTRID, LEGENDLABEL, NAME, NOMISSINGGROUP, TRANSPARENCY, X2AXIS, Y2AXIS

Related Style Elements: The GraphDataDefault style element is used to draw the plot elements. When a GROUP variable is present, or when multiple plot statements are overlaid, the GraphData1-GraphData12 elements are used, one for each group or plot statement.

Figure 4.15.1: Parametric Line

This graph shows city vs. highway mileage for all cars with a reference line of slope=1 to indicate same performance between city and Highway.

The 45° sloped reference line is drawn using the LINEPARM statement. A label for the line is displayed outside the data area using the CURVELABEL and CURVELABELLOC options.

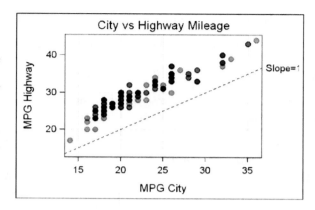

```
title 'City vs Highway Mileage';
proc sgplot data=sgbook.cars2;
   where type='Sedan';
   scatter x=mpgc y=mpgh / transparency=0.5 markerattrs=(symbol=circlefilled);
   lineparm x=0 y=0 slope=1.0 / curvelabel='Slope=1'
           curvelabelloc=outside lineattrs=(pattern=dash);
   xaxis grid;
   yaxis grid; run;
```

Figure 4.15.2: Parametric Lines

In this graph, the data for the two reference lines comes from the data set.

Three new columns "X", "Y" and "Slope" are merged into the data set. These columns are provided to the LINEPARM statement. The CURVELABEL and CURVELABELPOS options are used to display the labels.

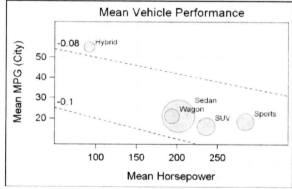

```
title 'Mean Vehicle Performance';
proc sgplot data=sgbook.carMeansLine;
   where _type_=1 and type<>'Truck';
   bubble x=mean_horsepower y=mean_mpg_city size=_freq_ /
          transparency=0.5 datalabel=type;
   lineparm x=x y=y slope=slope / curvelabel lineattrs=(pattern=dash)
          curvelabelpos=min;
   xaxis grid  offsetmin=0.15;
   yaxis grid;      run;
```

4.16 Waterfall Chart (9.3)

The WATERFALL statement has the following syntax:

```
waterfall category=column response=column </ options>;
```

The waterfall chart draws bar segments positioned relative to the cumulative sum of the preceding observations. The response value for each observation is expected to be a delta that is accumulated along the axis. The initial value for the accumulation can be specified. The final bar value is computed from the data.

Required Data Roles:
CATEGORY	column	Position along the horizontal axis
RESPONSE	num-column	Delta response along the vertical axis

Optional Data Roles:
COLORGROUP	=column	Classification for bar color
DATALABEL	=column	Label string for each observation
URL	=column	URL for each observation

Appearance Options:
DATALABELATTRS	=text-attrs	Appearance attributes of the labels
FILLATTRS	=fill-attrs	Appearance attributes of bubble interior
FINALBARATTRS	=line-attrs	Appearance attributes for bubble outline
INITIALBARATTRS	=line-attrs	Appearance attributes for bubble outline

Other Options:
BARWIDTH	=value	Radius in pixels of the largest bubble
DATASKIN	=skin-value	One of the predefined skin types
FINALBARTICKVALUE	=string	String for labeling final bar on axis
INITIALBARTICKVALUE	=string	String for labeling initial bar on axis
INITIALBARVALUE	=value	Response value for initial bar
STAT	=keyword	MEAN	SUM

Boolean Options:
FILL	NOFILL	Display filled bar or not
OUTLINE	NOOUTLINE	Display bar outline or not
MISSING	Show missing category values

Standard common options are supported
ATTRID, LEGENDLABEL, NAME, NOMISSINGGROUP, TRANSPARENCY, URL, X2AXIS, Y2AXIS

Related Style Elements: The GraphDataDefault style element is used to draw the plot elements. When a group variable is present, the GraphData1-GraphData12 elements are used, one for each group.

Figure 4.16.1: Basic Waterfall Chart

This graph shows the cumulative cash balance over transactions. Each observation is drawn relative to the previous cumulative statistic from zero, which is SUM in this case.

The final bar value is computed from the data and displayed at the end.

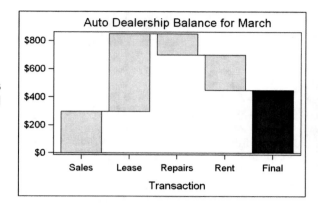

```
title "Auto Dealership Balance for March";
proc sgplot data=sgbook.transactions;
   waterfall category=transaction response=amount;
   yaxis display=(nolabel);
run;
```

Figure 4.16.2: Waterfall Chart with Initial Value

This graph shows the same data as the one shown in Figure 4.16.1 with the addition of setting the initial bar value. The cumulative statistic is computed starting with the initial value, which is 350 in this case.

The y-axis OFFSETMIN is set to zero and gridlines are enabled.

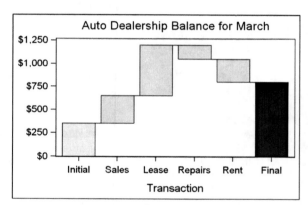

```
title "Auto Dealership Balance for March";
proc sgplot data=sgbook.transactions;
   waterfall category=transaction response=amount
             initialbarvalue=350;
   yaxis display=(nolabel) grid offsetmin=0;
   run;
```

Figure 4.16.3: Grouped Waterfall Chart

This graph shows the same data as the one shown in Figure 4.16.2 with the addition of a color group role and data skin. Here each bar is colored by the transaction type, Income or Expense.

A legend is displayed in the upper left corner of the graph.

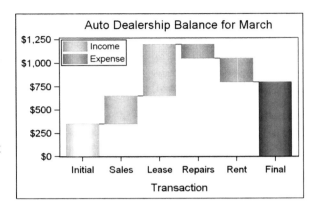

```
title "Auto Dealership Balance for March";
proc sgplot data=sgbook.transactions;
   waterfall category=transaction response=amount / colorgroup=Type
             initialbarvalue=350 initialbarattrs=(color=white)
             dataskin=pressed finalbarattrs=(color=cx2f2f2f);
   keylegend / location=inside position=topleft across=1;
   yaxis display=(nolabel) grid offsetmin=0;
   run;
```

Figure 4.16.4: Waterfall Chart with Data Labels

This graph shows the same data as the one shown in Figure 4.16.3 with the addition of data labels.

Long tick labels for initial and final bars is set, and XAXIS FITPOLICY is set to stagger.

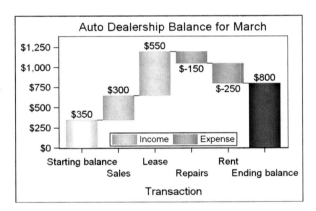

```
title "Auto Dealership Balance for March";
proc sgplot data=sgbook.transactions;
   waterfall category=transaction response=amount /
             initialbartickvalue="Starting balance"
             finalbartickvalue="Ending balance" datalabel
             initialbarattrs=(color=white) finalbarattrs=(color=black)
             colorgroup=Type dataskin=pressed initialbarvalue=350;
   keylegend / location=inside position=bottom;
   xaxis fitpolicy=stagger;
   yaxis display=(nolabel) offsetmin=0 grid;   run;
```

4.17 Combining the Plots

Basic plots can be combined (overlaid) in unlimited ways as long as the data types on each axis are consistent. In some of the preceding examples, we have seen various combinations of scatter, series, step, and band plots. These plot types tend to have numeric interval variables on both axes, but sometimes may have categorical variables on one or both axes.

The order of the plot statements and the axes determine if certain combinations are allowed. The axis type (linear, discrete, etc.) is decided by the first plot in the list. Subsequent plots must have compatible type of data on the same axis for the combination to be acceptable.

Compatibility:
1. If the axis is of type LINEAR, it is compatible with only LINEAR variables.
2. If the axis is of type DISCRETE, then it is compatible with both DISCRETE and LINEAR variables. Subsequent LINEAR data are treated as DISCRETE.

Axis and Data Compatibility:
1. If the first plot is a SCATTER, with x-axis having linear variables, then the default axis type is LINEAR. Now, subsequent overlaid plots must have linear variables for the combination to work. A plot with an x-axis that has a categorical variable will not be accepted, unless the axis type is explicitly set as DISCRETE.
2. If the first plot has a DISCRETE variable on the y-axis, then the subsequent overlaid plot can have a Y variable that is either linear or discrete. The linear data will be treated as DISCRETE.

Common Combinations of Basic Plots:
1. Any plot with one or more plots of the same kind – Multi-Response
2. Scatter with Series and Step
3. Series and Step plots with Band plots
4. Scatter, Series, and Bands
5. Scatter, Series, Step, Band, and BarParm with Reference Lines
6. Scatter, Series, Step, Band, and BarParm with LineParm Lines
7. Bar-Line using VBarParm, and Series
8. Scatter with MarkerChar - for overlaid annotated data labels
9. Scatter and Vector Plots
10. HighLow Type=Line with Series and Bands – Stock Plot
11. HighLow Type=Bar with Scatter – Adverse Event Plot

Note: Basic plot statements can also be combined with:
- Fit and Confidence plots – We will see these examples in Chapter 5.
- Distribution plots – We will see these examples in Chapter 6.
- Categorization plots – We will see these examples in Chapter 7.

Let us look at some examples in the following pages.

Figure 4.17.1: Scatter-Scatter Overlay

This is a combination of two scatter plots, one each for Min and Max Mileage (City) by mean Horsepower.

A legend is automatically generated, and we have positioned it in the upper right of the data area by using the KEYLEGEND statement.

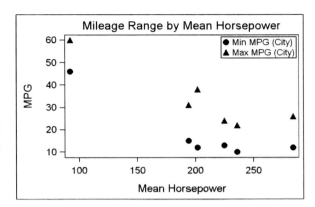

```
title 'Mileage Range by Mean Horsepower';
proc sgplot data=sgbook.carMeans;
  where _type_=1;
  scatter x=mean_horsepower y=min_mpg_city;
  scatter x=mean_horsepower y=max_mpg_city;
  keylegend / location=inside position=topright across=1;
  xaxis grid;
  yaxis grid label='MPG';
run;
```

Figure 4.17.2: Scatter-Series Overlay

This is an overlay of the raw data with a linear fit plot. The prediction curve has been computed previously using the REG procedure.

The standard GRAPHFIT style element is applied to the fit line.

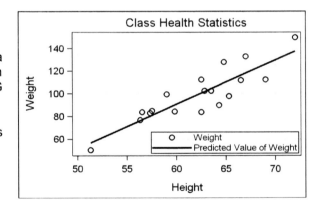

```
title 'Class Health Statistics';
proc sgplot data=sashelp.classfit;
  scatter x=height y=weight;
  series  x=height y=predict   / lineattrs=GraphFit;
  keylegend / location=inside position=bottomright across=1;
  xaxis grid;
  yaxis grid;
  run;
```

Figure 4.17.3: Scatter, Series, and Band

This is a standard fit plot with confidence limits. The prediction curve and the limits have been computed previously using the REG procedure.

The VALUEATTRS option on the KEYLEGEND statement is used to create smaller fonts.

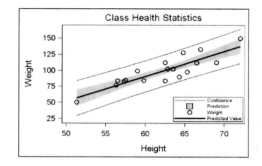

```
title 'Class Health Statistics';
proc sgplot data=sashelp.classfit;
  band x=height upper=upper lower=lower / legendlabel='Confidence' nofill;
  band x=height upper=uppermean lower=lowermean / legendlabel='Prediction';
  scatter x=height y=weight;
  series x=height y=predict / lineattrs=GraphFit legendlabel='Predicted Val';
  keylegend / location=inside position=bottomright across=1
              valueattrs=(size=6);
  xaxis grid;
  yaxis grid;
  run;
```

Figure 4.17.4: Step and Scatter Plot

Here is a survival plot by time and treatment. The statistics for the two treatments are in separate columns. This allows us to use the scatter plot to display the at-risk numbers.

A hidden REFLINE is used to label the at-risk numbers, using its "label" feature, while suppressing the line itself. Variable PlotD is used to suppress plotting of count.

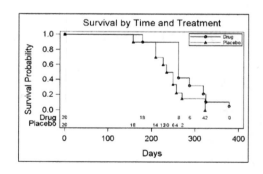

```
title 'Survival by Time and Treatment';
proc sgplot data=sgbook.survival;
  step x=daysD y=survivalD / markers markerattrs=GraphData1(size=4)
       name='D' legendlabel='Drug';
  step x=daysP y=survivalP / markers markerattrs=GraphData2(size=4)
       name='P' legendlabel='Placebo';
  scatter x=daysD y=ylabelD / markerchar=leftD markercharattrs=(size=3) freq=plotD;
  scatter x=daysP y=ylabelP / markerchar=leftP markercharattrs=(size=3) freq=plotP;
  refline -0.1 -0.2 / label=('Drug' 'Placebo') labelloc=outside labelpos=min
         lineattrs=(thickness=0);
  keylegend 'D' 'P' / location=inside position=topright across=1 valueattrs=(size=5);
  xaxis grid label='Days';    yaxis grid values=(0 to 1 by .2) valueshint;
  run;
```

Figure 4.17.5: Series and Band

This is a Body Mass Index graph, showing the regions that are classified as "Normal", "Underweight", etc. using a band plot.

The actual data values for each subject are plotted over time using a SERIES statement with markers and data labels. The CURVELABEL and CURVELABELPOS options are used to display curve labels.

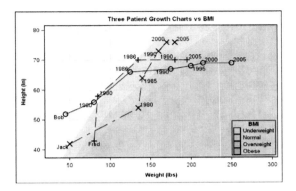

```
title 'Three Patient Growth Charts vs BMI';
proc sgplot data=sgbook.bmi;
  band x=weight_bmi upper=upper_hgt lower=lower_hgt /
       transparency=.7 group=category name="b";
  series x=wgt y=hgt / datalabel=year lineattrs=(thickness=1) markers
         group=name name="pts" curvelabel curvelabelpos=start;
  keylegend "b" / position=bottomright location=inside across=1 title='BMI';
  xaxis min=30  max=300  label='Weight (lbs)' grid;
  yaxis min=40  max=80   label='Height (in)'  grid;
run;
```

Figure 4.17.6: Scatter and Band

Here is an overlay of two band plots and a needle plot showing the sales figures over the year, with lower and upper limits.

The upper limits of each band are labeled, thus avoiding the need for a legend.

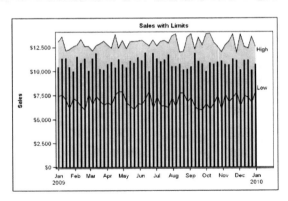

```
title 'Sales with Limits';
proc sgplot data=sgbook.sales noautolegend;
  band x=date upper=high lower=low / fillattrs=graphdata1
       fill outline curvelabelupper='High';
  band x=date upper=low lower=0 / fillattrs=graphdata5
       fill outline curvelabelupper='Low';
  needle x=date y=sales / lineattrs=(thickness=2) transparency=0.5;
  xaxis grid display=(nolabel);
  yaxis grid;
run;
```

Figure 4.17.7: Grouped Band

This is an example of a Theme River plot created using a grouped band plot. The data for each band covers the full height of the band. So, we have used TRANSPARENCY to ensure all bands are visible.

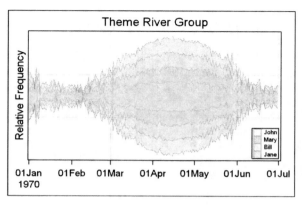

```
title 'Theme River Group';
proc sgplot data=sgbook.themeriverGroup;
  band x=time upper=upper lower=lower / group=name transparency=0.7
       fill outline lineattrs=(pattern=solid);
  keylegend / position=bottomright location=inside across=1
       valueattrs=(size=4);
  xaxis display=(nolabel) grid;
  yaxis grid label='Relative Frequency' display=(novalues noticks);
run;
```

Figure 4.17.8: Band Overlay

This graph is similar to Figure 4.17.7. Here the data for each band is in separate columns. So, we have used a four band overlay to create this graph.

Since we know the data extents of each band, we can specify the sequence for drawing, ensuring the narrowest band is drawn last. The TRANSPARENCY option is not used in this example.

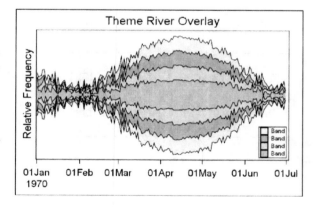

```
title 'Theme River Overlay';
proc sgplot data=sgbook.themeriverMulti;
  band x=time upper=JaneU lower=JaneL / fill outline lineattrs=(pattern=solid);
  band x=time upper=BillU lower=BillL / fill outline lineattrs=(pattern=solid);
  band x=time upper=MaryU lower=MaryL / fill outline lineattrs=(pattern=solid);
  band x=time upper=JohnU lower=JohnL / fill outline lineattrs=(pattern=solid);
  keylegend / position=bottomright location=inside across=1 valueattrs=(size=4);
  xaxis display=(nolabel) grid;
  yaxis grid label='Relative Frequency' display=(novalues noticks);
run;
```

Figure 4.17.9: Pareto Chart

Here is a Pareto Chart, using VBARPARM and SERIES statements.

BARWIDTH=0.8 and DISCRETEOFFSET= 0.4 are set to ensure the cumulative line is located at the top right corner of the bars.

Markers are displayed for the series plot, with data labels for each observation.

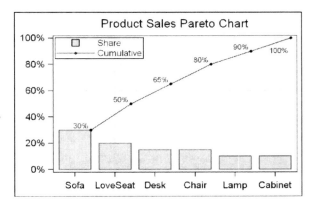

```
title 'Product Sales Pareto Chart';
proc sgplot data=sgbook.pareto;
  vbarparm category=product response=share / barwidth=0.8;
  series x=product y=cumulative / markers discreteoffset=0.4
      markerattrs=(symbol=circlefilled size=3) datalabel=cumulative;
  keylegend / position=topleft location=inside across=1;
  xaxis display=(nolabel) grid;
  yaxis grid label='Relative Frequency' display=(nolabel);
  run;
```

Figure 4.17.10: Floating Bars and Bands

This is a graph of the monthly average temperatures for a year in the future (2109), compared with 30-year average numbers.

The 30-year averages are potted using a HighLow plot with TYPE=Bar. The high and low data values are displayed. Values for 2109 are shown using a band plot.

VALUEATTRS is used for small fonts.

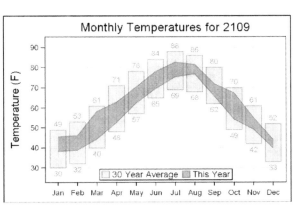

```
title 'Monthly Temperatures for 2109';
proc sgplot data=sgbook.weatherLine;
  highlow x=month high=high low=low / highlabel=high lowlabel=low
        type=bar legendlabel='30 Year Average' transparency=0.5;
  band x=month upper=thisyearH lower=thisyearL / fill outline
        transparency=0.5 fillattrs=graphdata3 legendlabel='This Year';
  keylegend / position=bottom location=inside;
  xaxis display=(nolabel) grid valueattrs=(size=7);
  yaxis grid label='Temperature (F)' valueattrs=(size=7);
  run;
```

Figure 4.17.11: VBARPARM with Statistics

This is a graph showing product sales with low and high forecasts. Here we have displayed some statistics that are not included in the graph.

Axis offsets are used to create space for the statistics. We have drawn the stats using scatter plots with MARKERCHAR associated with the Y2 axis. The space for the Y2 data is restricted to the top 15%.

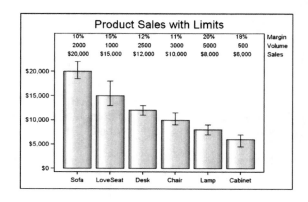

```
title 'Product Sales with Limits';
proc sgplot data=sgbook.productStat noautolegend;
  vbarparm category=product response=sales / dataskin=gloss
           limitupper=high limitlower=low;
  scatter x=product y=labelS / markerchar=sales y2axis;
  scatter x=product y=labelV / markerchar=vol y2axis;
  scatter x=product y=labelM / markerchar=margin y2axis;
  xaxis display=(nolabel) grid;
  yaxis grid display=(nolabel) offsetmax=0.2 values=(0 to 20000 by 5000) valueshint;
  y2axis offsetmin=0.85 display=(nolabel noticks);  run;
```

Figure 4.17.12: HBARPARM with Statistics

This graph is similar to Figure 4.17.11, using an HBarParm with limits, and scatter plots to display the statistics.

Axis offsets are used to create space for the statistics. We have drawn the stats using scatter plots with MARKERCHAR associated with the X2 axis. The space for the X2 data is restricted to the right 16%.

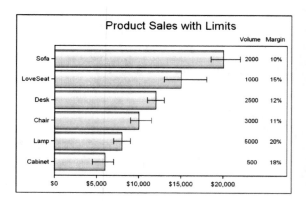

```
title 'Product Sales with Limits';
proc sgplot data=sgbook.productStat noautolegend;
  hbarparm category=product response=sales / dataskin=gloss
           limitupper=high limitlower=low;
  scatter y=product x=labelV / markerchar=vol x2axis;
  scatter y=product x=labelM / markerchar=margin x2axis;
  xaxis grid display=(nolabel) offsetmin=0 offsetmax=0.20
        values=(0 to 20000 by 5000) valueshint;
  x2axis offsetmin=0.84 offsetmax=0.05 display=(noticks nolabel) valueattrs=(size=7);
  yaxis display=(nolabel);  run;
```

Chapter 5

Fit and Confidence Plots

- 5.1 Introduction   129
- 5.2 Fit Plot Roles and Options   130
- 5.3 Regression Plot   131
- 5.4 Loess Plot   138
- 5.5 Penalized B-Spline Plot   145
- 5.6 Ellipse Plot   152
- 5.7 Combining the Plots   157

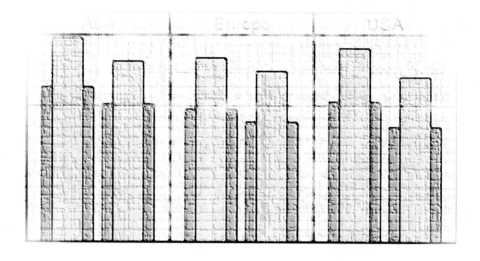

"Do, or do not. There is no 'try'."
~ Yoda

Chapter 5: Fit and Confidence Plots

In this chapter, we examine the fit and confidence plots for SGPLOT and SGPANEL procedures. The basic syntax for each plot and related options will be covered in detail in this chapter. Plots within this group can be mixed and matched with each other and with many X-Y plots from the "Basic Plots" group.

5.1 Introduction

The SGPLOT and SGPANEL procedures support multiple fit and confidence plots as listed below. These plots normally also display the raw data as a scatter plot, and also support display of confidence bands. These plots can be combined with many basic X-Y plots, most commonly with the scatter plot.

REG	LOESS
PBSPLINE	ELLIPSE

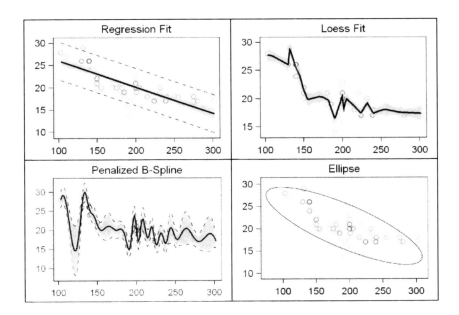

5.2 Fit Plot Roles and Options

All of the fit plot statements have the following syntax:

```
FitPlotName required-data-roles </ options>;
```

All plot statements have required and optional parameters necessary to create the visual. These parameters fall into the following broad categories:

1. required data roles
2. optional data roles
3. options
4. common options

Here is a list of common options for fits:

Common Options:

ATTRID	=string		The attribute map ID used for the group variable.
CURVELABELATTRS	=line-attrs		Appearance attributes for the curve label.
CURVELABELLOC	=keyword		Curve label location - INSIDE | OUTSIDE
CURVELABELPOS	=keyword		Curve label position - AUTO | MAX | MIN
DATALABEL	<=column>		Display a scatter label for the Y variable or from an optional variable.
DATALABELATTRS	=text-attrs		Appearance attributes for DataLabel.
GROUP	=column		Classification variable for fit calculations.
LEGENDLABEL	=string		The label that appears in the legend to represent this (non-group) plot.
LINEATTRS	= line-attrs		Appearance attributes for the fit line.
MARKERATTRS	= marker-attrs		Appearance attributes for the markers.
MAXPOINTS	=positive-integer		Maximum number of prediction points.
NAME	=string		Specifies a name for this statement. Other statements, such as a KEYLEGEND, can refer to a plot by its name.
TRANSPARENCY	=value		Specifies the transparency for the visual elements.

Common Boolean Options:

NOMARKERS	Do not show the scatter markers with the fit line.
X2AXIS	Assigns X values to the X2 (top) axis.
Y2AXIS	Assigns Y values to the Y2 (right) axis.

These common options are available for each plot statement, and can be used exactly as described above. To avoid duplication, we will not list the above common options for the individual plots that are discussed in this chapter.

5.3 Regression Plot

The Regression plot shows a linear or nonlinear fit of the raw data. The syntax is:

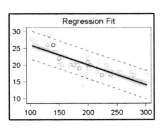

```
reg x=column y=column </ options>;
```

A marker is displayed at each (x, y) location. A fit line is computed and displayed. Optionally, different confidence limits can also be displayed.

Required Data Roles:

X	=num-column	Position along the horizontal axis
Y	=num-column	Position along the vertical axis

Optional Data Roles:

FREQ	=num-column	Frequency of the current observation
WEIGHT	=num-column	Priority weights for a regression fit

Appearance Options:

ALPHA	=positive-number	Confidence level
CLIATTRS	=cli-attrs	Appearance attributes for CLI
CLMATTRS	=clm-attrs	Appearance attributes for CLM
DEGREE	=positive-integer	The degree of the polynomial

Other Options:

CLI	=string	Display the confidence limits for individual predicted values using the string for the legend label.
CLM	=string	Display confidence limits for mean predicted values using the string for the legend label.

Boolean Options:

CLI	Display confidence limits for individual predicted values.
CLM	Display confidence limits for mean predicted values.
NOLEGCLI	Exclude the CLI from the legend.
NOLEGCLM	Exclude the CLM from the legend.
NOLEGFIT	Exclude the fit line from the legend.

Standard common options are supported by fits
ATTRID, CURVELABEL, CURVELABELLOC, CURVELABELPOS, DATALABEL, DATALABELATTRS, GROUP, LEGENDLABEL, LINEATTRS, MARKERATTRS, MAXPOINTS, NAME, NOMARKERS, TRANSPARENCY, X2AXIS, Y2AXIS

Related Style Elements: The GraphDataDefault style element is used to draw the plot elements. When a GROUP variable is present, or when multiple plot statements are overlaid, the GraphData1-GraphData12 elements are used, one for each group or plot statement.

Figure 5.3.1: Basic Linear Fit

This plot contains a basic linear regression fit. By default, the REG statement produces both the fit line and the scatter points. This behavior can be controlled by the NOMARKERS option.

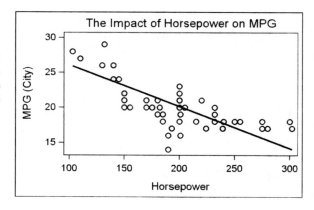

```
title "The Impact of Horsepower on MPG";
proc sgplot data=sashelp.cars noautolegend;
  where type='Sedan' and origin='USA';
  reg x=horsepower y=mpg_city;
  xaxis grid;
  yaxis grid;
  run;
```

Figure 5.3.2: CLM Band

In addition to the linear fit and scatter points, this plot displays a computed confidence limit for the mean (CLM).

By default, the alpha value for the computation is .05 (95% confidence). This value can be adjusted using the ALPHA option.

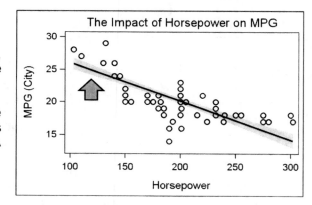

```
title "The Impact of Horsepower on MPG";
proc sgplot data=sashelp.cars noautolegend;
  where type='Sedan' and origin='USA';
  reg x=horsepower y=mpg_city / clm;
  xaxis grid;
  yaxis grid;
  run;
```

Figure 5.3.3: CLI Band

In addition to the linear fit, scatter points, and CLM band, this plot displays the prediction limits for individual values (CLI).

By default, the alpha value for the computation is .05 (95% confidence). This value can be adjusted using the ALPHA option, which affects both CLM and CLI.

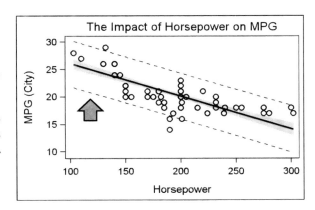

```
title "The Impact of Horsepower on MPG";
proc sgplot data=sashelp.cars noautolegend;
   where type='Sedan' and origin='USA';
   reg x=horsepower y=mpg_city / clm cli;
   xaxis grid;
   yaxis grid;
   run;
```

Figure 5.3.4: Automatic Legend

This plot displays a legend containing information about the fit. This legend appears by default and can be disabled by using the NOAUTOLEGEND option or by specifying a custom legend.

The NOLEGFIT option prevents the fit line information from being displayed in the legend.

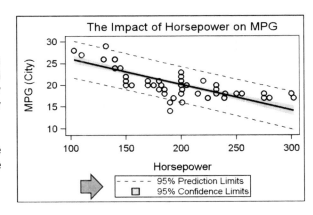

```
title "The Impact of Horsepower on MPG";
proc sgplot data=sashelp.cars;
   where type='Sedan' and origin='USA';
   reg x=horsepower y=mpg_city / clm cli nolegfit;
   xaxis grid;
   yaxis grid;
   run;
```

Figure 5.3.5: Axis Assignment

In this plot, the regression fit is assigned to both the X and Y secondary axes.

Notice that the grid line must now be specified on the X2AXIS and Y2AXIS statements.

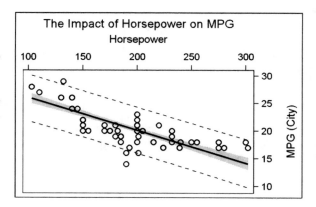

```
title "The Impact of Horsepower on MPG";
proc sgplot data=sashelp.cars noautolegend;
  where type='Sedan' and origin='USA';
  reg x=horsepower y=mpg_city / clm cli x2axis y2axis;
  x2axis grid;
  y2axis grid;
  run;
```

Figure 5.3.6: Attribute Control

Many of the visual attributes of the REG statement can be controlled with options.

In this example, the CLI line is set to a "solid" pattern and the scatter marker shapes are set to "plus".

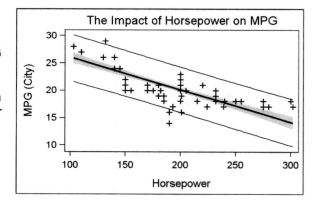

```
title "The Impact of Horsepower on MPG";
proc sgplot data=sashelp.cars noautolegend;
  where type='Sedan' and origin='USA';
  reg x=horsepower y=mpg_city / clm cli markerattrs=(symbol=plus)
      cliattrs=(clilineattrs=(pattern=solid)) ;
  xaxis grid;
  yaxis grid;
  run;
```

Figure 5.3.7: Scatter Data Labels

By default, the DATALABEL option displays the non-missing Y values of the scatter points. You can assign a column to the option to display alternate values.

In this example, the Y values of only the outlying points were copied to a column called "outlier". The missing values in that column are not displayed.

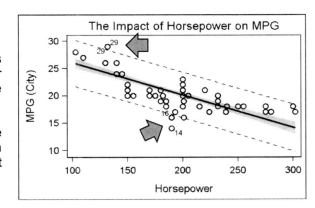

```
title "The Impact of Horsepower on MPG";
proc sgplot data=sgbook.carlabel noautolegend;
  where type='Sedan' and origin='USA';
  reg x=horsepower y=mpg_city / clm cli datalabel=outlier;
  xaxis grid;
  yaxis grid;
  run;
```

Figure 5.3.8: Polynomial Regression Fit

The DEGREE option can be used to change the polynomial for the regression fit.

In this example, a quadratic fit is made through the scatter points by setting the DEGREE option to 2.

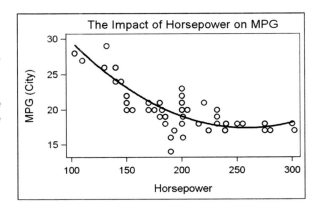

```
title "The Impact of Horsepower on MPG";
proc sgplot data=sashelp.cars noautolegend;
  where type='Sedan' and origin='USA';
  reg x=horsepower y=mpg_city / degree=2;
  xaxis grid;
  yaxis grid;
  run;
```

Figure 5.3.9: Grouped Fits

When a GROUP column is specified on a REG statement, both the scatter points and the fit lines are grouped.

Two separate statements are used to control the transparency of the markers. The NOMARKER option on the REG statement is used to disable the default scatter points.

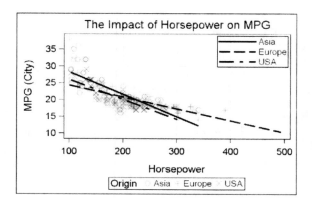

```
title "The Impact of Horsepower on MPG";
proc sgplot data=sashelp.cars;
  where type='Sedan';
  scatter x=horsepower y=mpg_city / group=origin transparency=0.85
          name="scat";
  reg x=horsepower y=mpg_city / group=origin nomarkers name="reg";
  keylegend "scat" / title="Origin";
  keylegend "reg" / location=inside across=1;
  xaxis grid;
  yaxis grid; run;
```

Figure 5.3.10: Single Fit with a Group

To create a single fit through all of your grouped scatter data, use a SCATTER statement to plot the grouped data and a REG statement without a group column and the NOMARKER option.

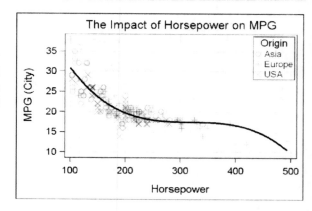

```
title "The Impact of Horsepower on MPG";
proc sgplot data=sashelp.cars;
  where type='Sedan';
  scatter x=horsepower y=mpg_city / group=origin transparency=0.85
          name="scat";
  reg x=horsepower y=mpg_city / nomarkers degree=3 name="reg";
  keylegend "scat" / location=inside across=1 title="Origin";
  xaxis grid;
  yaxis grid;
  run;
```

Figure 5.3.11: Overlaid Fits

You can overlay fits for comparison by using the NOMARKER option on all REG statements but one, or you can use a separate SCATTER statement for more control over the scatter plot.

Notice that the REG statements are assigned to a custom legend. The LEGENDLABEL option overrides the default legend label for each fit.

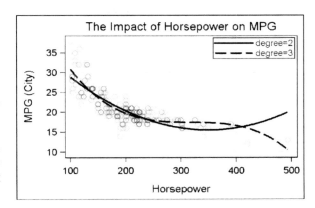

```
title "The Impact of Horsepower on MPG";
proc sgplot data=sashelp.cars;
  where type='Sedan';
  scatter x=horsepower y=mpg_city / transparency=0.85;
  reg x=horsepower y=mpg_city / nomarkers degree=2 name="reg1"
      legendlabel="degree=2";
  reg x=horsepower y=mpg_city / nomarkers degree=3 name="reg2"
      legendlabel="degree=3";
  keylegend "reg1" "reg2" / location=inside across=1;
  xaxis grid;  yaxis grid; run;
```

Figure 5.3.12: Curve Labels

Instead of using a legend to identify each fit, you can use the CURVELABEL option to directly label the fit lines. This approach works well when the lines are not too crowded.

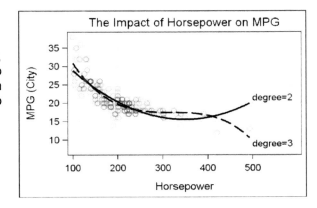

```
title "The Impact of Horsepower on MPG";
proc sgplot data=sashelp.cars noautolegend;
  where type='Sedan';
  scatter x=horsepower y=mpg_city / transparency=0.85;
  reg x=horsepower y=mpg_city / nomarkers degree=2 curvelabel="degree=2";
  reg x=horsepower y=mpg_city / nomarkers degree=3 curvelabel="degree=3";
  xaxis grid;
  yaxis grid;
  run;
```

5.4 Loess Plot

The Loess plot shows a smoothed fit of the raw data with a Loess fit. The syntax is:

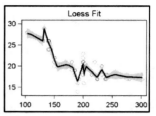

```
loess x=column y=column </ options>;
```

A marker is displayed at each (x, y) location. A fit line is computed and displayed. Optionally, confidence limits of the means can be displayed.

Required Data Roles:

X	=num-column	Position along the horizontal axis.
Y	=num-column	Position along the vertical axis.

Optional Data Roles:

WEIGHT	=num-column	Priority weights for a regression fit.

Appearance Options:

ALPHA	=positive-number	Confidence level to compute.
CLMATTRS	=clm-attrs	Appearance attributes for CLM.
DEGREE	=positive-integer	The degree of the polynomial.
INTERPOLATION	=keyword	The degree of the interpolating polynomials used for blending local polynomial fits at the kd tree vertices – LINEAR | CUBIC.
REWEIGHT	=postive-integer	The number of iterative reweighting steps.
SMOOTH	=postive-integer	Specifies a smoothing parameter value.

Other Options:

CLM	=string	Display confidence limits for mean predicted values using the string for the legend label.

Boolean Options:

CLM	Display confidence limits for mean predicted values.
NOLEGCLM	Exclude the CLM from the legend.
NOLEGFIT	Exclude the fit line from the legend.

Standard common options are supported by fits

ATTRID, CURVELABEL, CURVELABELLOC, CURVELABELPOS, DATALABEL, DATALABELATTRS, GROUP, LEGENDLABEL, LINEATTRS, MARKERATTRS, MAXPOINTS, NAME, NOMARKERS, TRANSPARENCY, X2AXIS, Y2AXIS

Related Style Elements: The GraphDataDefault style element is used to draw the plot elements. When a GROUP variable is present, or when multiple plot statements are overlaid, the GraphData1-GraphData12 elements are used, one for each group or plot statement.

Figure 5.4.1: Basic Loess Fit

This plot contains a basic loess fit. By default, the LOESS statement produces both the fit line and the scatter points. This behavior can be controlled by the NOMARKERS option.

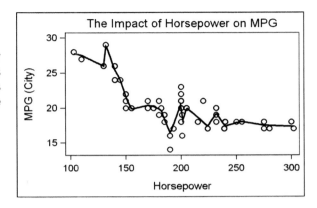

```
title "The Impact of Horsepower on MPG";
proc sgplot data=sashelp.cars noautolegend;
   where type='Sedan' and origin='USA';
   loess x=horsepower y=mpg_city;
   xaxis grid;
   yaxis grid;
   run;
```

Figure 5.4.2: Fit Plot with Smooth Curves

The SMOOTH option enables you to specify a smoothing parameter for the loess fit.

Compare the default fit of this data in Figure 5.4.1 with this output.

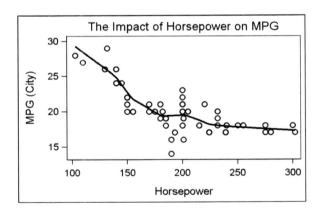

```
title "The Impact of Horsepower on MPG";
proc sgplot data=sashelp.cars noautolegend;
   where type='Sedan' and origin='USA';
   loess x=horsepower y=mpg_city / smooth=0.4;
   xaxis grid;
   yaxis grid;
   run;
```

Figure 5.4.3: CLM Band

In addition to the linear fit and scatter points, this plot displays the confidence limit of the mean (CLM).

By default, the alpha value for the computation is .05 (95% confidence). This value can be adjusted using the ALPHA option.

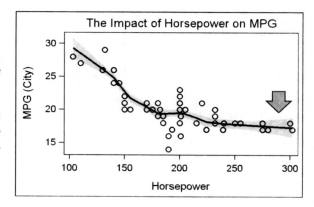

```
title "The Impact of Horsepower on MPG";
proc sgplot data=sashelp.cars noautolegend;
  where type='Sedan' and origin='USA';
  loess x=horsepower y=mpg_city / smooth=0.4 clm;
  xaxis grid;
  yaxis grid;
  run;
```

Figure 5.4.4: Automatic Legend

This plot displays a legend containing information about the fit. This legend appears by default and can be disabled by using the NOAUTOLEGEND option or by specifying a custom legend.

The NOLEGFIT option prevents the fit line information from being displayed in the legend.

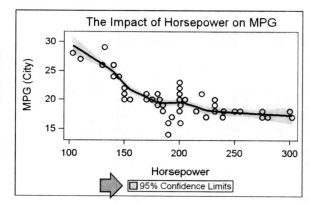

```
title "The Impact of Horsepower on MPG";
proc sgplot data=sashelp.cars;
  where type='Sedan' and origin='USA';
  loess x=horsepower y=mpg_city / smooth=0.4 clm nolegfit;
  xaxis grid;
  yaxis grid;
  run;
```

Figure 5.4.5: Axis Assignment

In this plot, the loess fit is assigned to both the X and Y secondary axes.

Notice that the grid line must now be specified on the X2AXIS and Y2AXIS statements.

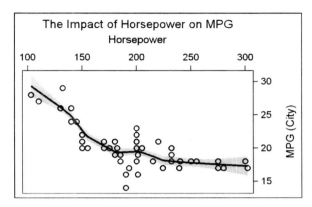

```
title "The Impact of Horsepower on MPG";
proc sgplot data=sashelp.cars noautolegend;
  where type='Sedan' and origin='USA';
  loess x=horsepower y=mpg_city / smooth=0.4 clm x2axis y2axis;
  x2axis grid;
  y2axis grid;
  run;
```

Figure 5.4.6: Attribute Control

Many of the visual attributes of the LOESS statement can be controlled with options.

In this example, the CLM band is set to a "#dddddd" gray color, and the scatter marker shapes are set to "plus".

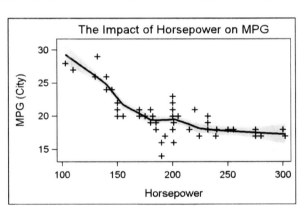

```
title "The Impact of Horsepower on MPG";
proc sgplot data=sashelp.cars noautolegend;
  where type='Sedan' and origin='USA';
  loess x=horsepower y=mpg_city / clm markerattrs=(symbol=plus)
        clmattrs=(clmfillattrs=(color=graydd)) smooth=0.4;
  xaxis grid;
  yaxis grid;
  run;
```

Figure 5.4.7: Scatter Data Labels

By default, the DATALABEL option displays the non-missing Y values of the scatter points. You can assign a column to the option to display alternate values.

In this example, the Y values of the min/max points were copied to a column called "minmax". The missing values in that column are not displayed.

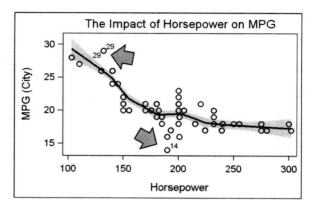

```
title "The Impact of Horsepower on MPG";
proc sgplot data=sgbook.carlabel2 noautolegend;
   where type='Sedan' and origin='USA';
   loess x=horsepower y=mpg_city / clm smooth=0.4 datalabel=minmax;
   xaxis grid;
   yaxis grid;
   run;
```

Figure 5.4.8: Cubic Loess Fit

The INTERPOLATION option can be used to change the degree of the interpolating polynomials. In this example, a cubic fit is made through the scatter points by setting the INTERPOLATION option to CUBIC.

Compare this output to the default LINEAR interpolation output in Figure 5.4.1.

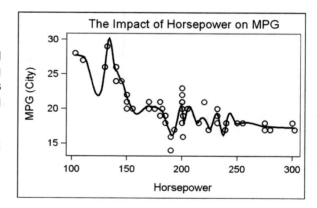

```
title "The Impact of Horsepower on MPG";
proc sgplot data=sashelp.cars noautolegend;
   where type='Sedan' and origin='USA';
   loess x=horsepower y=mpg_city / interpolation=cubic;
   xaxis grid;
   yaxis grid;
   run;
```

Figure 5.4.9: Grouped Fits

When a GROUP column is specified on a LOESS statement, both the scatter points and the fit lines are grouped.

Two separate statements are used to control the transparency of the markers. The NOMARKERS option on the LOESS statement is used to disable the default scatter points.

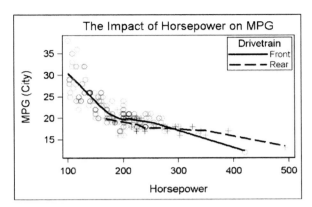

```
title "The Impact of Horsepower on MPG";
proc sgplot data=sashelp.cars;
  where type='Sedan' and drivetrain ne "All";
  scatter x=horsepower y=mpg_city / group=DriveTrain transparency=0.8;
  loess x=horsepower y=mpg_city / group=DriveTrain smooth=0.4
        nomarkers name="loess";
  keylegend "loess" / location=inside across=1 title="Drivetrain";
  xaxis grid;
  yaxis grid;
  run;
```

Figure 5.4.10: Single Fit with a Group

To create a single fit through all of your grouped scatter data, use a SCATTER statement to plot the grouped data and a LOESS statement without a group column and the NOMARKERS option.

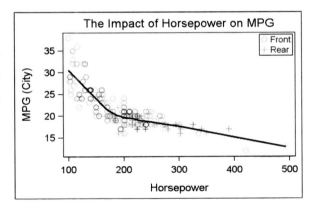

```
title "The Impact of Horsepower on MPG";
proc sgplot data=sashelp.cars;
  where type='Sedan' and drivetrain ne "All";
  scatter x=horsepower y=mpg_city / group=DriveTrain
          transparency=0.8 name="scat";
  loess x=horsepower y=mpg_city /  smooth=0.4 nomarkers;
  keylegend "scat" / location=inside across=1;
  xaxis grid;
  yaxis grid; run;
```

Figure 5.4.11: Overlaid Fits

You can overlay fits for comparison by using the NOMARKERS option on all LOESS statements but one, or you can use a separate SCATTER statement for more control over the scatter plot.

Notice that the LOESS statements are assigned to a custom legend. The default legend label reflects the fit type used and the smoothing parameter specified.

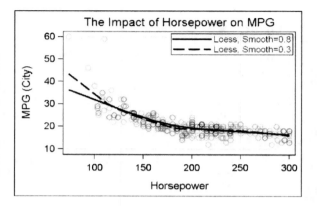

```
title "The Impact of Horsepower on MPG";
proc sgplot data=sashelp.cars;
   where horsepower <= 300;
   scatter x=horsepower y=mpg_city / transparency=0.8;
   loess x=horsepower y=mpg_city / nomarkers smooth=0.8 name="loess1";
   loess x=horsepower y=mpg_city / nomarkers smooth=0.3 name="loess2";
   keylegend "loess1" "loess2" / location=inside across=1;
   xaxis grid;
   yaxis grid; run;
```

Figure 5.4.12: Curve Labels

Instead of using a legend to identify each fit, you can use the CURVELABEL option to directly label the fit lines. This approach works well when the lines are not too crowded.

Here the labels are moved to the MIN side of the fits and outside of the data area. The plots are assigned to the secondary y-axis to give space to the fit labels on the left.

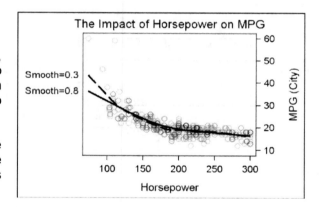

```
title "The Impact of Horsepower on MPG";
proc sgplot data=sashelp.cars noautolegend;
   where horsepower <= 300;
   scatter x=horsepower y=mpg_city / transparency=0.8 y2axis;
   loess x=horsepower y=mpg_city / nomarkers smooth=0.8
      curvelabel="Smooth=0.8" curvelabelpos=min curvelabelloc=outside y2axis;
   loess x=horsepower y=mpg_city / nomarkers smooth=0.3
      curvelabel="Smooth=0.3" curvelabelpos=min curvelabelloc=outside y2axis;
   xaxis grid;
   y2axis grid;   run;
```

5.5 Penalized B-Spline Plot

The penalized B-spline plot shows a smoothed fit of the raw data created with a penalized B-spline:

```
pbspline x=column y=column </ options>;
```

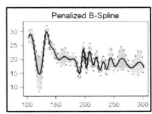

A marker is displayed at each (x, y) location. A fit line is computed and displayed. Optionally, confidence limits and/or confidence limits of the means can be displayed.

Required Data Roles:
X	=num-column	Position along the horizontal axis
Y	=num-column	Position along the vertical axis

Optional Data Roles:
FREQ	=num-column	Frequency of the current observation
WEIGHT	=num-column	Priority weights for a regression fit

Appearance Options:
ALPHA	=positive-number	Confidence level to compute
CLIATTRS	=cli-attrs	Appearance attributes for CLI
CLMATTRS	=clm-attrs	Appearance attributes for CLM
DEGREE	=positive-integer	The degree of the polynomial
NKNOTS	=positive-number	Number of evenly-spaced internal knots

Other Options:
CLI	=string	Display confidence limits for individual predicted values using the string for the legend label.
CLM	=string	Display confidence limits for mean predicted values using the string for the legend label.
CLMTRANSPARENCY	=value	Transparency for the CLM band.

Boolean Options:
CLI	Display confidence limits for individual predicted values.
CLM	Display confidence limits for mean predicted values.
NOLEGCLI, NOLEGCLM	Exclude the CLI or CLM from the legend.
NOLEGFIT	Exclude the fit line from the legend.

Standard common options are supported by fits
ATTRID, CURVELABEL, CURVELABELLOC, CURVELABELPOS, DATALABEL, DATALABELATTRS, GROUP, LEGENDLABEL, LINEATTRS, MARKERATTRS, MAXPOINTS, NAME, NOMARKERS, X2AXIS, Y2AXIS

Related Style Elements: The GraphDataDefault style element is used to draw the plot elements. When a GROUP variable is present, or when multiple statements are overlaid, the GraphData1-GraphData12 elements are used, one for each group or statement.

Figure 5.5.1: Basic Parametric B-Spline Fit

This plot contains a basic penalized B-spline fit. By default, the PBSPLINE statement produces both the fit line and the scatter points. This behavior can be controlled by the NOMARKERS option.

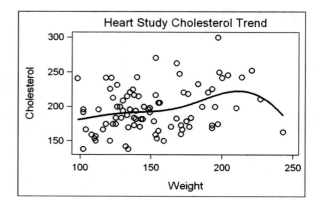

```
title "Heart Study Cholesterol Trend";
proc sgplot data=sashelp.heart noautolegend;
  where ageatstart <= 30;
  pbspline x=weight y=cholesterol;
  xaxis grid;
  yaxis grid;
  run;
```

Figure 5.5.2: CLM Band

In addition to the PBSpline fit and scatter points, this plot displays a computed confidence limit of the mean (CLM).

By default, the alpha value for the computation is .05 (95% confidence). This value can be adjusted using the ALPHA option.

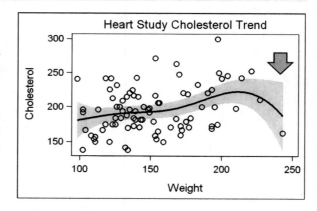

```
title "Heart Study Cholesterol Trend";
proc sgplot data=sashelp.heart noautolegend;
  where ageatstart <= 30;
  pbspline x=weight y=cholesterol / clm;
  xaxis grid;
  yaxis grid;
  run;
```

Figure 5.5.3: CLI Band

In addition to the PBSpline fit, scatter points, and CLM band, this plot displays the confidence limits (CLI) for individual predicted values for each observation.

By default, the alpha value for the computation is .05 (95% confidence). This value can be adjusted using the ALPHA option, which affects both CLM and CLI.

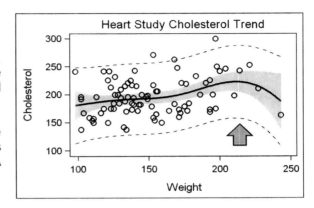

```
title "Heart Study Cholesterol Trend";
proc sgplot data=sashelp.heart noautolegend;
   where ageatstart <= 30;
   pbspline x=weight y=cholesterol / clm cli;
   xaxis grid;
   yaxis grid;
   run;
```

Figure 5.5.4: Automatic Legend

This plot displays a legend containing information about the fit. This legend appears by default and can be disabled by using the NOAUTOLEGEND option or by specifying a custom legend.

The NOLEGFIT option prevents the fit line information from being displayed in the legend.

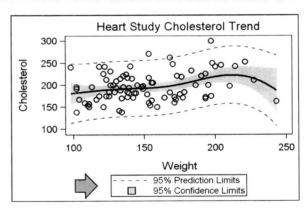

```
title "Heart Study Cholesterol Trend";
proc sgplot data=sashelp.heart;
   where ageatstart <= 30;
   pbspline x=weight y=cholesterol / clm cli nolegfit;
   xaxis grid;
   yaxis grid;
   run;
```

Figure 5.5.5: Axis Assignment

In this plot, the PBSPLINE fit is assigned to both the X and Y secondary axes.

Notice that the grid lines must now be specified on the X2AXIS and Y2AXIS statements.

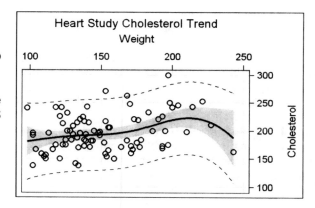

```
title "Heart Study Cholesterol Trend";
proc sgplot data=sashelp.heart noautolegend;
  where ageatstart <= 30;
  pbspline x=weight y=cholesterol / clm cli x2axis y2axis;
  x2axis grid;
  y2axis grid;
  run;
```

Figure 5.5.6: Attribute Control

Many of the visual attributes of the PBSPLINE statement can be controlled with options.

In this example, the CLI line is set to a "solid" pattern and the scatter marker shapes are set to "plus".

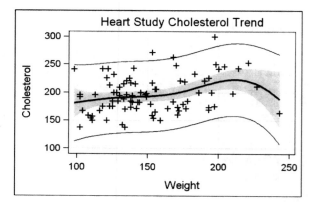

```
title "Heart Study Cholesterol Trend";
proc sgplot data=sashelp.heart noautolegend;
  where ageatstart <= 30;
  pbspline x=weight y=cholesterol / clm cli
         cliattrs=(clilineattrs=(pattern=solid))
         markerattrs=(symbol=plus);
  xaxis grid;
  yaxis grid;
  run;
```

Figure 5.5.7: Scatter Data Labels

By default, the DATALABEL option displays the non-missing Y values of the scatter points. You can assign a column to the option to display alternate values.

In this example, the Y values of the outlying points were copied to a column called "outlier". The missing values in that column are not displayed.

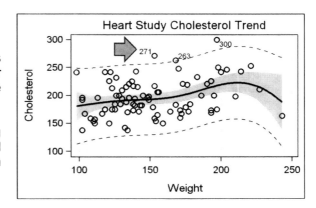

```
title "Heart Study Cholesterol Trend";
proc sgplot data=sgbook.heartLabel noautolegend;
  where ageatstart <= 30;
  pbspline x=weight y=cholesterol / clm cli datalabel=outlier;
  xaxis grid;
  yaxis grid;
  run;
```

Figure 5.5.8: The DEGREE Option

The DEGREE option can be used to change the degree of the B-spline.

In this example, the DEGREE option is set to 1. Compare the result with the default result in Figure 5.5.1, where the degree is 3.

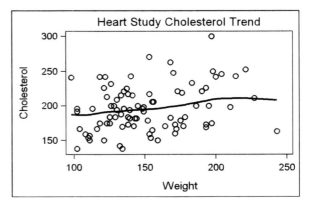

```
title "Heart Study Cholesterol Trend";
proc sgplot data=sashelp.heart noautolegend;
  where ageatstart <= 30;
  pbspline x=weight y=cholesterol / degree=1;
  xaxis grid;
  yaxis grid;
  run;
```

Figure 5.5.9: Grouped Fits

When a GROUP column is specified on a PBSPLINE statement, both the scatter points and the fit lines are grouped.

Two separate statements are used to control the transparency of the markers. The NOMARKER option on the PBSPLINE statement is used to disable the default scatter points.

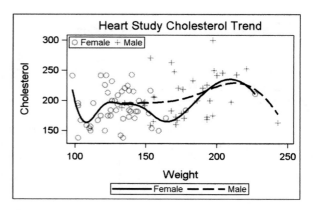

```
title "Heart Study Cholesterol Trend";
proc sgplot data=sashelp.heart;
  where ageatstart <= 30;
  scatter x=weight y=cholesterol / group=sex transparency=0.5 name="scat";
  pbspline x=weight y=cholesterol / group=sex nomarkers name="fit";
  keylegend "fit";
  keylegend "scat" / location=inside;
  xaxis grid;
  yaxis grid;
  run;
```

Figure 5.5.10: Single Fit with a Group

To create a single fit through all of your grouped scatter data, use a SCATTER statement to plot the grouped data and a PBSPLINE statement without a group column and the NOMARKER option.

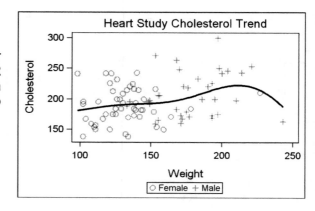

```
title "Heart Study Cholesterol Trend";
proc sgplot data=sashelp.heart;
  where ageatstart <= 30;
  scatter x=weight y=cholesterol / group=sex transparency=0.5 name="scat";
  pbspline x=weight y=cholesterol / nomarkers;
  keylegend "scat";
  xaxis grid;
  yaxis grid;
  run;
```

Figure 5.5.11: Overlaid Fits

You can overlay fits for comparison by using the NOMARKERS option on all PBSPLINE statements but one, or you can use a separate SCATTER statement for more control over the scatter plot.

Notice that the PBSPLINE statements are assigned to a custom legend. The LEGENDLABEL option overrides the default legend label for each fit.

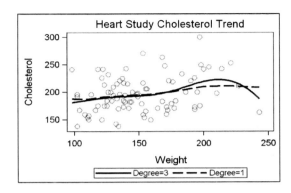

```
title "Heart Study Cholesterol Trend";
proc sgplot data=sashelp.heart;
  where ageatstart <= 30;
  scatter x=weight y=cholesterol / transparency=0.5;
  pbspline x=weight y=cholesterol / nomarkers degree=3 name="three"
           legendlabel="Degree=3";
  pbspline x=weight y=cholesterol / nomarkers degree=1 name="one"
           legendlabel="Degree=1";
  keylegend "three" "one";
  xaxis grid;
  yaxis grid;
  run;
```

Figure 5.5.12: Curve Labels

Instead of using a legend to identify each fit, you can use the CURVELABEL option to directly label the fit lines.

This approach works well when the lines are not too crowded.

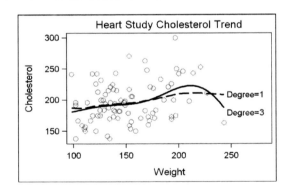

```
title "Heart Study Cholesterol Trend";
proc sgplot data=sashelp.heart noautolegend;
  where ageatstart <= 30;
  scatter x=weight y=cholesterol / transparency=0.5;
  pbspline x=weight y=cholesterol / nomarkers degree=3 curvelabel="Degree=3";
  pbspline x=weight y=cholesterol / nomarkers degree=1 curvelabel="Degree=1";
  xaxis grid;
  yaxis grid;
  run;
```

5.6 Ellipse Plot

The Ellipse plot shows a confidence ellipse of the raw data. The syntax is:

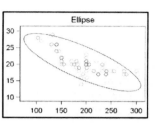

```
ellipse x=column y=column </ options>;
```

A marker is displayed at each (x, y) location. An ellipse is computed and displayed.

Required Data Roles:

X	=*num-column*	Position along the horizontal axis
Y	=*num-column*	Position along the vertical axis

Optional Data Roles:

FREQ	=*num-column*	Frequency of the current observation

Appearance Options:

ALPHA	=*positive-number*	Confidence level to compute
FILLATTRS	=*fill-attrs*	Attributes for the ellipse fill
LINEATTRS	=*line-attrs*	Attributes for the ellipse outline
TYPE	=*keyword*	Type of the confidence ellipse PREDICTED | MEAN

Boolean Options:

CLIP	Exclude the ellipse data from the axis data range calculation
FILL / NOFILL	Enable/Disable display of a fill color in the ellipse
OUTLINE/NOOUTLINE	Enable/Disable display of an outline around the ellipse

Standard common options are supported by fits
  LEGENDLABEL, NAME, TRANSPARENCY, X2AXIS, Y2AXIS

Related Style Elements: The GraphDataDefault style element is used to draw the plot elements. When a GROUP variable is present, or when multiple plot statements are overlaid, the GraphData1-GraphData12 elements are used, one for each group or plot statement.

Figure 5.6.1: Basic Ellipse

By default, the ELLIPSE statement produces a prediction ellipse with an ALPHA value of .05.

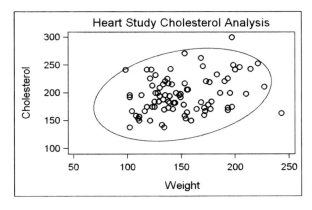

```
title "Heart Study Cholesterol Analysis";
proc sgplot data=sashelp.heart noautolegend;
  where ageatstart <= 30;
  ellipse x=weight y=cholesterol;
  scatter x=weight y=cholesterol;
  xaxis grid;
  yaxis grid;
  run;
```

Figure 5.6.2: The Filled Ellipse

When the FILL option is specified, the ellipse is filled with color and the outline is not displayed.

The outline can be re-enabled by specifying the OUTLINE option.

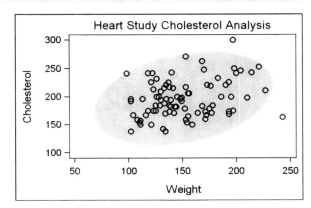

```
title "Heart Study Cholesterol Analysis";
proc sgplot data=sashelp.heart noautolegend;
  where ageatstart <= 30;
  ellipse x=weight y=cholesterol / fill;
  scatter x=weight y=cholesterol;
  xaxis grid;
  yaxis grid;
  run;
```

Figure 5.6.3: Ellipse Clipping

The CLIP option prevents the ellipse data from contributing to the axis range, causing the ellipse to clip if it extends beyond the other data.

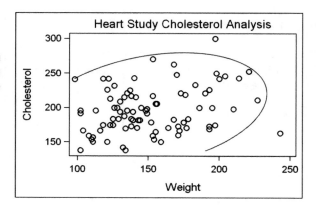

```
title "Heart Study Cholesterol Analysis";
proc sgplot data=sashelp.heart noautolegend;
  where ageatstart <= 30;
  ellipse x=weight y=cholesterol / clip;
  scatter x=weight y=cholesterol;
  xaxis grid;
  yaxis grid;
  run;
```

Figure 5.6.4: Ellipse with Confidence Level

The ALPHA option sets the confidence level for the computation. By default, the alpha value is .05.

Compare this output with Figure 5.6.1 to see the effect of the alpha change.

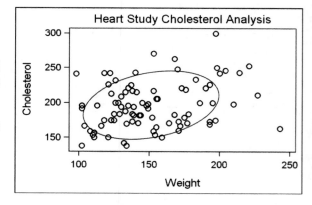

```
title "Heart Study Cholesterol Analysis";
proc sgplot data=sashelp.heart noautolegend;
  where ageatstart <= 30;
  ellipse x=weight y=cholesterol / alpha=.35;
  scatter x=weight y=cholesterol;
  xaxis grid;
  yaxis grid;
  run;
```

Figure 5.6.5: Ellipse Attributes

When the fill or outline features are visible, the attributes can be controlled from the ELLIPSE statement.

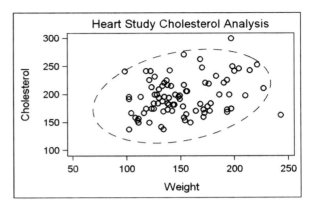

```
title "Heart Study Cholesterol Analysis";
proc sgplot data=sashelp.heart noautolegend;
  where ageatstart <= 30;
  ellipse x=weight y=cholesterol / fill outline lineattrs=(pattern=dash)
        fillattrs=(color=cxeeeeee);
  scatter x=weight y=cholesterol;
  xaxis grid;
  yaxis grid;
  run;
```

Figure 5.6.6: Ellipse Type

By default, the ellipse type is PREDICTED, which creates a prediction ellipse for a new observation.

In this example, the TYPE is set to MEAN, which specifies a confidence ellipse of the population mean.

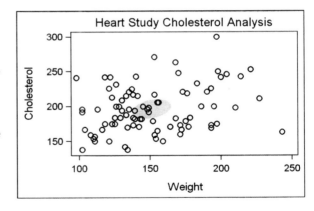

```
title "Heart Study Cholesterol Analysis";
proc sgplot data=sashelp.heart noautolegend;
  where ageatstart <= 30;
  ellipse x=weight y=cholesterol / type=mean fill alpha=.001;
  scatter x=weight y=cholesterol;
  xaxis grid;
  yaxis grid;
  run;
```

Figure 5.6.7: Overlaid Ellipses

In this example, three ellipses with different alpha values are overlaid and associated with a legend. The legend reflects the information about the ellipses.

In addition, the cholesterol values for the outlying points are copied to a column called "outlier" and displayed using the DATALABEL option.

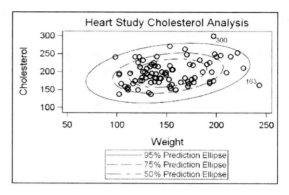

```
title "Heart Study Cholesterol Analysis";
proc sgplot data=sgbook.heartlabel2 noautolegend;
  where ageatstart <= 30;
  ellipse x=weight y=cholesterol / alpha=.05 name="e1";
  ellipse x=weight y=cholesterol / alpha=.25 name="e2";
  ellipse x=weight y=cholesterol / alpha=.50 name="e3";
  scatter x=weight y=cholesterol / datalabel=outlier;
  keylegend "e1" "e2" "e3";
  xaxis grid;
  yaxis grid;  run;
```

Figure 5.6.8: Overlaid Filled Ellipses

This example is similar to Figure 5.6.7, except that the ellipses are filled, creating a backdrop for the scatterplot.

The ellipse representation in the legend also changes with the presence of the FILL option.

Ellipses are ordered in the program so that the smaller ellipses are drawn on top.

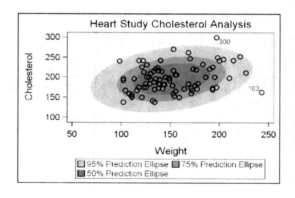

```
title "Heart Study Cholesterol Analysis";
proc sgplot data= sgbook.heartlabel2 noautolegend;
  where ageatstart <= 30;
  ellipse x=weight y=cholesterol / fill alpha=.05 name="e1";
  ellipse x=weight y=cholesterol / fill alpha=.25 name="e2";
  ellipse x=weight y=cholesterol / fill alpha=.50 name="e3";
  scatter x=weight y=cholesterol / datalabel=outlier;
  keylegend "e1" "e2" "e3";
  xaxis grid;
  yaxis grid;  run;
```

5.7 Combining the Plots

Figure 5.7.1: Comparing Fits

Different types of fits with various parameters can be overlaid on the raw plot data for comparison.

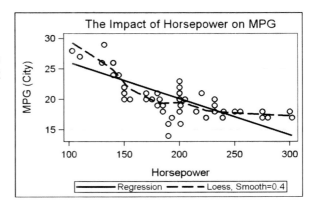

```
title "The Impact of Horsepower on MPG";
proc sgplot data=sashelp.cars;
  where type='Sedan' and origin='USA';
  reg x=horsepower y=mpg_city;
  loess x=horsepower y=mpg_city / nomarkers smooth=0.4;
  xaxis grid;
  yaxis grid;  run;
```

Figure 5.7.2: Comparing Fits with CLM

To prevent the CLM bands from overlapping the fit lines, the statements are ordered to draw the CLM bands first, then the fit lines.

The first two plot statements suppress the fit lines by specifying a zero-thickness line attribute. The legend is assigned to the statements that draw the fit lines.

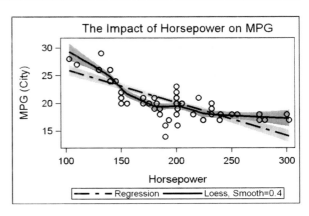

```
proc sgplot data=sashelp.cars;
  where type='Sedan' and origin='USA';
  reg x=horsepower y=mpg_city / clm nomarkers lineattrs=(thickness=0);
  loess x=horsepower y=mpg_city / clm nomarkers smooth=0.4
        lineattrs=(thickness=0);
  reg x=horsepower y=mpg_city / nomarkers name="reg";
  loess x=horsepower y=mpg_city / name="loess" smooth=0.4
        lineattrs=(pattern=solid);
  keylegend "reg" "loess";
  xaxis grid;  yaxis grid;  run;
```

Chapter 6

Distribution Plots

6.1 Introduction 161
6.2 Histogram 162
6.3 Density Plot 166
6.4 Vertical Box Plot 169
6.5 Horizontal Box Plot 174
6.6 Combining the Plots 177

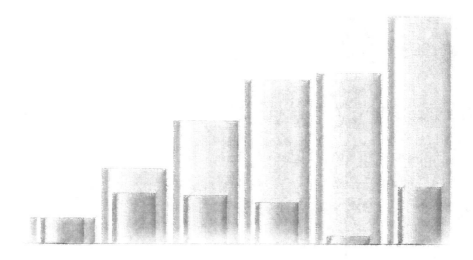

"Information is not knowledge."
~ Albert Einstein

Chapter 6: Distribution Plots

In this chapter, we examine the distribution plots that are supported by the SGPLOT and SGPANEL procedures. The basic syntax for each plot and the options will be covered in detail in this chapter. Plots within this group can be mixed and matched with each other and with many X-Y plots from the basic plots group.

6.1 Introduction

The SGPLOT and SGPANEL procedures support multiple distribution plots as listed below. These plots normally do not display the raw data. These plots can be combined only in specific ways with other plots in this category or the basic plots.

HISTOGRAM	DENSITY
VBOX	HBOX

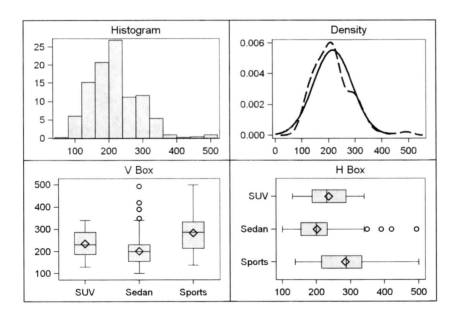

6.2 Histogram

The histogram shows the distribution of a numeric variable. The syntax is:

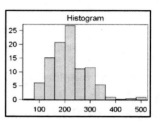

```
histogram column </ options>;
```

The observations are binned into N number of bins determined by the statement. The counts in the bin are shown as percentage by default. Alternatively, frequency counts can be displayed.

Required Data Roles:

X	num-column	Position along the horizontal axis

Optional Data Roles:

FREQ	=num-column	Frequency count for each observation

Appearance Options:

FILLATTRS	=fill-attrs	Appearance attributes for bins

Other Options:

BINSTART	= value	X coordinate of the first bin.
BINWIDTH	=value	Width of the bin
BOUNDARY	=keyword	Location of boundary values – UPPER | LOWER
NBINS	=value	Number of bins
SCALE	=keyword	PERCENT | COUNT | PROPORTION for Y axis scaling

Boolean Options:

FILL | NOFILL	Whether or not the bins are filled
OUTLINE | NOOUTLINE	Whether or not the bins have outlines
SHOWBINS	Tickmark on the X axis is shown at the midpoint of the bin

Standard common options are supported

LEGENDLABEL, NAME, TRANSPARENCY, X2AXIS, Y2AXIS

Related Style Elements: The GraphDataDefault style element is used to draw the plot elements.

Figure 6.2.1: Basic Histogram

This is a basic histogram of horsepower for only the sedans in the data set.

Default scale for y-axis is percent. The x-axis label is suppressed.

By default, the x-axis is linear.

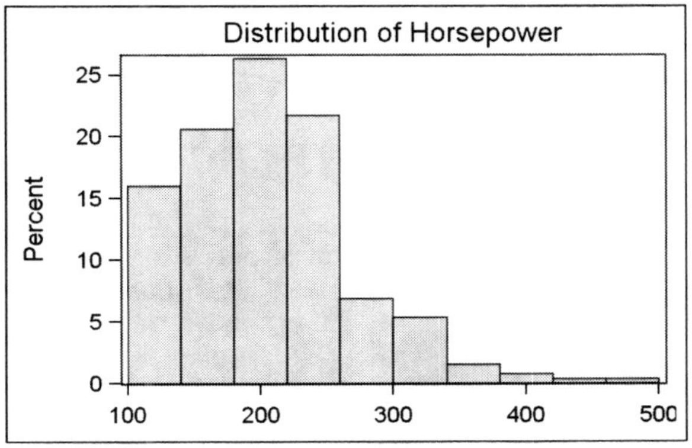

```
title 'Distribution of Horsepower';
proc sgplot data=sashelp.cars;
   where type='Sedan';
   histogram horsepower;
   yaxis  grid;
   xaxis display=(nolabel);
   run;
```

Figure 6.2.2: Histogram with Frequency

This is a histogram of horsepower for the cars. The FREQ role is used to apply the frequency for each observation. The data set itself only has observations where type='Sedan'.

In this graph the y-axis is "Count". The x-axis label is suppressed.

By default, the x-axis is linear.

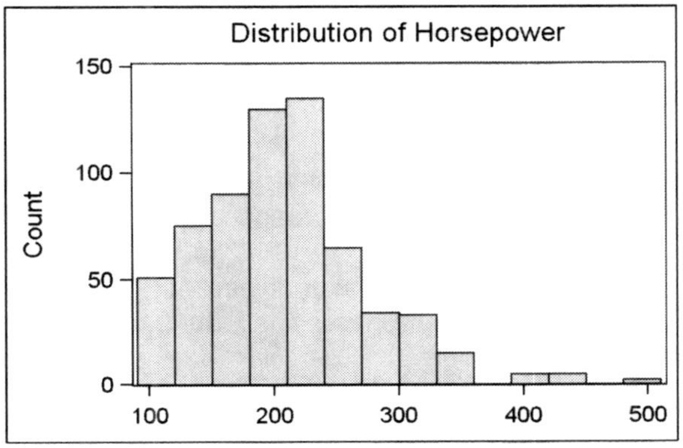

```
title 'Distribution of Horsepower';
proc sgplot data=sgbook.carsFreq;
   histogram horsepower / freq=freq scale=count;
   yaxis  grid;
   xaxis display=(nolabel);
   run;
```

Figure 6.2.3: Histogram with NBINS

For this histogram, the NBINS=25. This is a hint for the plot; the actual bins may be different.

Default scale for y-axis is percent. The x-axis label is suppressed. Transparency is set to 40%.

By default, the x-axis is linear.

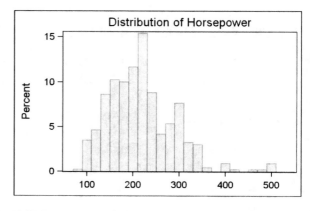

```
title 'Distribution of Horsepower';
proc sgplot data=sashelp.cars;
   histogram horsepower / nbins=25 transparency=0.4;
   yaxis  grid;
   xaxis display=(nolabel);
   run;
```

Figure 6.2.4: Histogram with BINSTART and BINWIDTH

For this histogram the BINSTART=0 and the BINWIDTH=20. Bins are created starting from zero with a width of 20 until the max data value is crossed.

Default scale for y-axis is percent. The x-axis label is suppressed. Transparency is set to 40%.

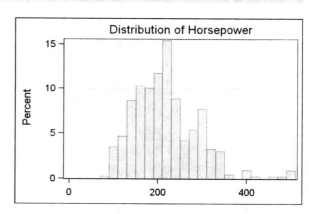

```
title 'Distribution of Horsepower';
proc sgplot data=sashelp.cars;
   histogram horsepower / binstart=0 binwidth=20 transparency=0.4;
   yaxis  grid;
   xaxis display=(nolabel);
   run;
```

Figure 6.2.5: Multiple Histograms

This graph shows multiple histograms, one for city and one for highway mileage.

For one of the plots we have set NOFILL to distinguish the two, and see through to the first. Color and transparency can also be used to distinguish the two histograms.

A default legend is automatically created.

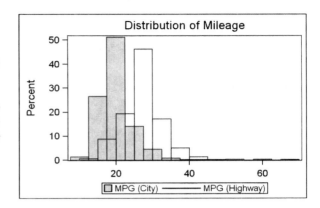

```
title 'Distribution of Mileage';
proc sgplot data=sashelp.cars;
   histogram mpg_City / transparency=0.5;
   histogram mpg_highway / nofill;
   yaxis  grid;
   xaxis display=(nolabel);
   run;
```

Figure 6.2.6: Multiple Histograms

This graph shows multiple histograms, one each for diastolic and systolic blood pressure. BINSTART and BINWIDTH are set on both histograms for similar binning.

Reference lines show the normal ranges. Transparency is used to distinguish between the histograms.

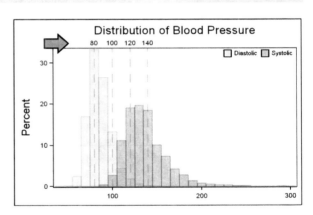

```
title 'Distribution of Blood Pressure';
proc sgplot data=sashelp.heart;
   histogram diastolic / binstart=40 binwidth=10 transparency=0.7;
   histogram systolic / binstart=40 binwidth=10 transparency=0.5;
   refline 80 100 120 140 / axis=X label=('80' '100' '120' '140')
           labelpos=max lineattrs=(pattern=dash);
   keylegend / location=inside position=topright noborder;
   xaxis display=(nolabel);
   yaxis   grid;  run;
```

6.3 Density Plot

The density plot shows the fitted distribution of a numeric variable.
The syntax is:

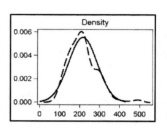

```
density column </ options>;
```

The distribution of the observations is shown as a curve. The plot supports normal density and a kernel density estimate.

Required Data Roles:

	num-column	Position along the horizontal axis

Optional Data Roles:

FREQ	=*num-column*	Frequency count for each observation

Appearance Options:

LINEATTRS	=*line-attrs*	Appearance attributes for line

Other Options:

TYPE	=*keyword (opts)*	NORMAL (options) | KENNEL (options)
SCALE	=*keyword*	PERCENT | COUNT | PROPORTION for y-axis scaling

Type=Normal Options:

MU	=*value*	Mean value to be used
SIGMA	=*value*	Standard deviation to be used

Type=Kernel Options:

C	=*value*	Band Width
WEIGHT	=*keyword*	NORMAL | QUADRATIC | TRIANGULAR

Standard common options are supported

LEGENDLABEL, NAME, TRANSPARENCY, X2AXIS, Y2AXIS

Related Style Elements: The GraphDataDefault style element is used to draw the plot elements.

Figure 6.3.1: Normal Density Curve

This is a graph of the normal density distribution for horsepower for all cars in the data set.

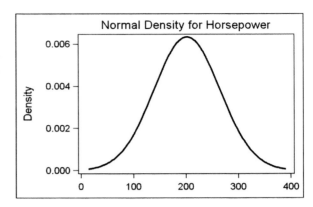

```
title 'Normal Density for Horsepower';
proc sgplot data=sashelp.cars noautolegend;
   where type='Sedan';
   density horsepower;
   yaxis grid;
   xaxis display=(nolabel);
   run;
```

Figure 6.3.2: Density Curve Options

This graph is of the same data as in Figure 6.3.1, with SCALE=Percent, and using a dash line pattern.

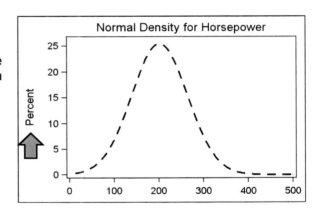

```
title 'Normal Density for Horsepower';
proc sgplot data=sashelp.cars noautolegend;
   where type='Sedan';
   density horsepower / scale=percent lineattrs=(pattern=dash);
   yaxis grid;
   xaxis display=(nolabel);
   run;
```

Figure 6.3.3: Normal Density Curve

The normal density function can be customized by specifying the Mu and Sigma parameters.

In this graph SIGMA=30, as shown in the inset.

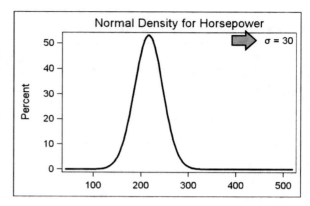

```
ods escapechar '^';
title 'Normal Density for Horsepower';
proc sgplot data=sashelp.cars noautolegend;
  density horsepower / scale=percent type=Normal(sigma=30);
  inset "^{unicode SIGMA} = 30" / position=topright;
  yaxis  grid;
  xaxis display=(nolabel);
  run;
```

Figure 6.3.4: Kernel Density Estimate

This graph shows the kernel density estimate for horsepower.

TYPE=Kernel supports options for C and weight.

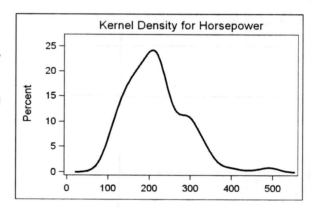

```
title 'Kernel Density for Horsepower';
proc sgplot data=sashelp.cars noautolegend;
  density horsepower / scale=percent type=Kernel;
  yaxis  grid;
  xaxis display=(nolabel);
  run;
```

6.4 Vertical Box Plot

The vertical box (VBOX) plot shows the distribution of the analysis variable by category. The syntax is:

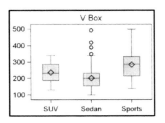

```
vbox num-column </ options>;
```

A box is displayed for each unique category value. The default display includes the Q1-Q3 interval, mean, median, whiskers, and outliers.

Required Data Roles:
	num-column	Analysis variable.

Optional Data Roles:
CATEGORY	=column	For category values along x-axis
FREQ	=num-column	Frequency count for each observation
GROUP	=column	Group variable for each observation

Appearance Options:
CONNECTATTRS	=line-attrs	Appearance for connect line
DATALABELATTRS	=text-attrs	Appearance for data labels
FILLATTRS	=fill-attrs	Appearance of box interior
LINEATTRS	=line-attrs	Appearance of box outlines
MEANATTRS OUTLIERATTRS	=marker-attrs	Appearance of mean and outliers
MEDIANATTRS	=line-attrs	Appearance of median
WHISKERATTRS	=line-attrs	Appearance of whiskers

Other Options:
BOXWIDTH	=value	Width of the box as a fraction of spacing
CAPSHAPE	=keyword	SERIF	LINE	BRACKET
CONNECT	=keyword	The statistic to connect – MEAN, MEDIAN, etc.

Boolean Options:
DATALABEL	Label the outliers
EXTREME	Whiskers are drawn to the extreme values
FILL	NOFILL	Specifies whether the boxes are filled or not
LABELFAR	Display labels for far outliers
MISSING NOCAPS NOMEAN	Various options
NOMEDIAN NOOUTLIERS NOTCHES	Various options

Standard common options are supported
CLUSTERWIDTH, DISCRETEOFFSET, GROUPDISPLAY, GROUPORDER, LEGENDLABEL, NAME, NOMISSINGGROUP, TRANSPARENCY, X2AXIS, Y2AXIS

Related Style Elements: Same as for Scatter, Series, VBar, and HBar.

Figure 6.4.1: Basic Box Plot

This is a basic box plot of mileage by type. Default display includes the Q1-Q3 interval, mean marker, median line, whiskers, and outliers.

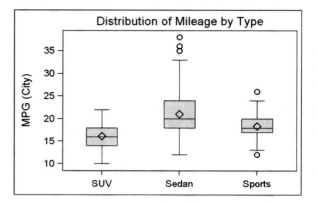

```
title 'Distribution of Mileage by Type';
proc sgplot data=sashelp.cars noautolegend;
  where type='Sedan' or type='Sports' or type='SUV';
  vbox mpg_city / category=type;
  yaxis  grid;
  xaxis display=(nolabel);
  run;
```

Figure 6.4.2: Grouped Box Plot

This is a grouped box plot of mileage by type and drivetrain. For consistency with other graphs like scatter, series, etc., the default GROUPDISPLAY is Overlay.

Set GROUPDISPLAY=Cluster to get side-by-side display of group values.

A legend is displayed by default.

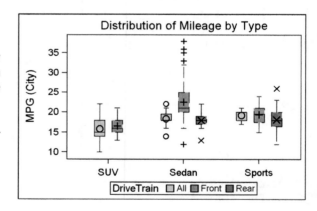

```
title 'Distribution of Mileage by Type';
proc sgplot data=sashelp.cars;
  where type='Sedan' or type='Sports' or type='SUV';
  vbox mpg_city / category=type group=drivetrain
       groupdisplay=cluster;
  yaxis  grid;
  xaxis display=(nolabel);
  run;
```

Figure 6.4.3: Grouped Unfilled Box Plot

This is a grouped box plot, where we have set the line patterns for the box to solid, and set NOFILL.

The legend displays the line patterns for the whiskers to distinguish the boxes.

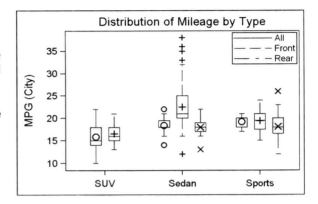

```
title 'Distribution of Mileage by Type';
proc sgplot data=sashelp.cars;
  where type='Sedan' or type='Sports' or type='SUV';
  vbox mpg_city / category=type group=drivetrain groupdisplay=cluster
                  nofill lineattrs=(pattern=solid);
  keylegend / location=inside position=topright across=1;
  yaxis  grid;
  xaxis display=(nolabel);
run;
```

Figure 6.4.4: Box Plot Overlay

This graph shows the distribution of mileage for city and highway as an overlay of two separate variables. The two box plots use DISCRETEOFFSET and BOXWIDTH to display the boxes side by side.

With SAS 9.3, box plots with common category axis can be overlaid. This is a common use case in clinical trials for multiple treatments.

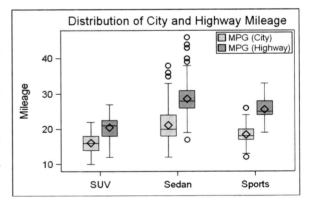

```
title 'Distribution of City and Highway Mileage';
proc sgplot data=sashelp.cars;
  where type='Sedan' or type='Sports' or type='SUV';
  vbox mpg_city / category=type discreteoffset=-0.12 boxwidth=0.2;
  vbox mpg_highway / category=type boxwidth=0.2 discreteoffset= 0.12;
  keylegend / location=inside position=topright across=1;
  yaxis  grid label='Mileage';
  xaxis display=(nolabel);
run;
```

Figure 6.4.5: Box Plot on Linear Axis

This is a graph of a VBOX where the x-axis data is set to linear (SAS 9.3) and so is not categorical. Format is MONNAME3.

This is a common use case in the health and life sciences industry, where the data is distributed unevenly on the time axis.

XAXIS TYPE=linear. The box spacing is determined by the smallest interval in data.

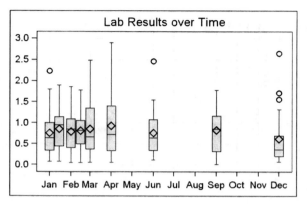

```
title 'Lab Results over Time';
proc sgplot data=sgbook.intervalBoxMulti;
  vbox DrugA / category=date;
  xaxis grid type=linear display=(nolabel) tickvalueformat=monname3
        values=('01Jan2009'd to '01Dec2009'd by month);
  yaxis grid display=(nolabel);
  run;
```

Figure 6.4.6: Box Overlay for Linear Data

This is an overlay of two VBOX plots on a date axis with uneven values.

Here we want to view the lab results over time and by treatment type. A VBOX overlay must have the same X variable, so we have to jitter the date values appropriately. See the code for generating this data set.

XAXIS TYPE=linear.

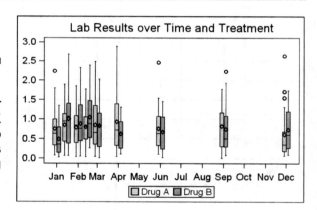

```
title 'Lab Results over Time and Treatment';
proc sgplot data=sgbook.intervalBoxMulti;
  vbox DrugA / category=date2 meanattrs=(size=5) outlierattrs=(size=5);
  vbox DrugB / category=date2 meanattrs=(size=5) outlierattrs=(size=5);
  xaxis grid type=linear display=(nolabel) offsetmin=0.05 offsetmax=0.05
        values=('01Jan2009'd to '01Dec2009'd by month)
        tickvalueformat=monname3.;
  yaxis grid display=(nolabel);
  run;
```

Figure 6.4.7: Box Plot with Labels

This graph shows a VBOX plot of mileage by vehicle type, with CONNECT=Mean.

DATALABEL option displays the data labels for the outliers.

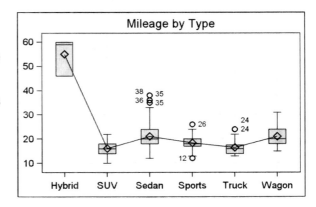

```
title 'Mileage by Type';
proc sgplot data=sashelp.cars;
  vbox mpg_city / category=type connect=mean datalabel;
  xaxis grid display=(nolabel);
  yaxis grid display=(nolabel);
run;
```

Figure 6.4.8: Box Plot with Notches

This is a graph showing boxes with notches using the NOTCHES option.

OFFSETMAX=0.2 is set to provide space for the notches in this case.

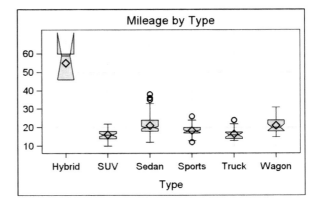

```
title 'Mileage by Type';
proc sgplot data=sashelp.cars;
  vbox mpg_city / category=type notches;
  xaxis grid;
  yaxis grid display=(nolabel) offsetmax=0.2;
run;
```

6.5 Horizontal Box Plot

The horizontal box (HBOX) plot shows the distribution of the analysis variable by category. The syntax is:

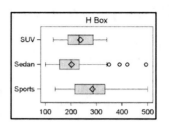

```
hbox column </ options>;
```

A box is displayed for each unique category value. The default display includes the Q1-Q3 interval, mean, median, whiskers, and outliers.

Required Data Roles:

	num-column	Analysis variable

Optional Data Roles:

CATEGORY	=*column*	For category values along x-axis
FREQ	=*num-column*	Frequency count for each observation
GROUP	=*column*	Group variable for each observation

Appearance Options:

CONNECTATTRS	=*line-attrs*	Appearance for connect line
DATALABELATTRS	=*text-attrs*	Appearance for data labels
FILLATTRS	=*attrs*	Appearance of box interior
LINEATTRS	=*attrs*	Appearance of box outlines
MEANATTRS OUTLIERATTRS	=*marker-attrs*	Appearance of mean and outliers
MEDIANATTRS	=*line-attrs*	Appearance of median
WHISKERATTRS	=*line-attrs*	Appearance of whiskers

Other Options:

BOXWIDTH	=*value*	Width of the box as a fraction of spacing
CAPSHAPE	=*keyword*	SERIF | LINE | BRACKET
CONNECT	=*keyword*	The statistic to connect – MEAN, MEDIAN, etc.

Boolean Options:

DATALABEL	Label the outliers
EXTREME	Whiskers are drawn to the extreme values
FILL | NOFILL	Specifies whether the boxes are filled or not
LABELFAR	Display labels for far outliers
MISSING NOCAPS NOMEAN	Various options
NOMEDIAN NOOUTLIERS NOTCHES	Various options

Standard common options are supported
  CLUSTERWIDTH, DISCRETEOFFSET, GROUPDISPLAY, GROUPORDER, LEGENDLABEL, NAME, NOMISSINGGROUP, TRANSPARENCY, X2AXIS, Y2AXIS

Related Style Elements: Same as for VBOX.

Figure 6.5.1: Basic Horizontal Box Plot

This graph uses the basic HBOX to display the distribution of cholesterol by cause of death.

A horizontal orientation for the box plot is particularly useful for larger number of category values having long strings.

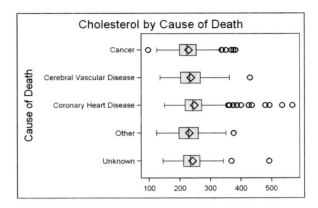

```
title 'Cholesterol by Cause of Death';
proc sgplot data=sashelp.heart;
  hbox cholesterol / category=deathcause;
  xaxis grid display=(nolabel);
  yaxis grid;
run;
```

Figure 6.5.2: Horizontal Box Plot with Notches

This graph uses the HBOX with NOTCHES to display the distribution of cholesterol by cause of death.

A horizontal orientation for the box plot is particularly useful where the category values have long character strings.

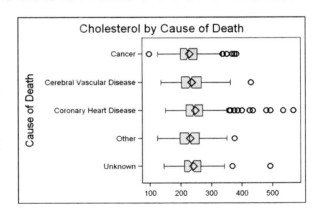

```
title 'Cholesterol by Cause of Death';
proc sgplot data=sashelp.heart;
  hbox cholesterol / category=deathcause notches;
  xaxis grid display=(nolabel);
  yaxis grid;
run;
```

Figure 6.5.3: Grouped Horizontal Box Plot

This graph shows the distribution of cholesterol by cause of death with GROUP=Sex.

LINEATTRS option sets the outline for all boxes to solid.

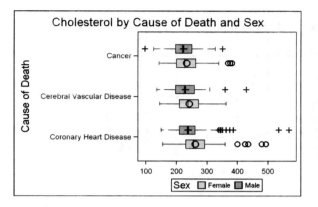

```
title 'Cholesterol by Cause of Death and Sex';
proc sgplot data=sashelp.heart;
  where deathcause <> 'Other' and deathcause <> 'Unknown';
  hbox cholesterol / category=deathcause group=sex
       lineattrs=(pattern=solid);
  xaxis grid display=(nolabel);
  yaxis grid;
  run;
```

Figure 6.5.4: Grouped Horizontal Box with CLUSTERWIDTH

This graph shows the distribution of cholesterol by cause of death with GROUP=Sex.

LINEATTRS option sets the outline for all boxes to solid. CLUSTERWIDTH=0.5 brings the group values closer.

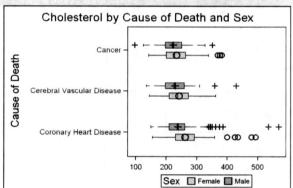

```
title 'Cholesterol by Cause of Death and Sex';
proc sgplot data=sashelp.heart;
  where deathcause <> 'Other' and deathcause <> 'Unknown';
  hbox cholesterol / category=deathcause group=sex
       lineattrs=(pattern=solid) clusterwidth=0.5;
  xaxis grid display=(nolabel);
  yaxis grid;
  run;
```

6.6 Combining the Plots

Figure 6.6.1: Histogram with Normal Density Plot

This graph displays the distribution of the mileage for all cars, along with the normal density curve.

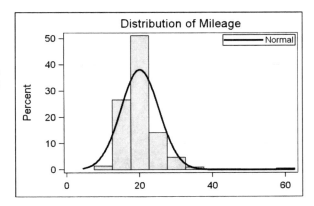

```
title 'Distribution of Mileage';
proc sgplot data=sashelp.cars;
   histogram mpg_city;
   density mpg_city;
   keylegend / location=inside position=topright;
   xaxis display=(nolabel);
   yaxis grid;    run;
```

Figure 6.6.2: Histogram with Normal and Kernel Density Plots

This graph displays the distribution of the mileage for all cars. The normal density curve and the kernel density estimate are included.

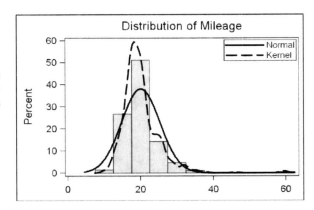

```
title 'Distribution of Mileage';
proc sgplot data=sashelp.cars;
   histogram mpg_city;
   density mpg_city;
   density mpg_city / type=kernel;
   keylegend / location=inside position=topright across=1;
   xaxis display=(nolabel);
   yaxis grid;    run;
```

Chapter 7

Categorization Plots

7.1   Introduction   181
7.2   Categorization Plot Roles and Common Options   182
7.3   Vertical Bar Charts   183
7.4   Horizontal Bar Charts   191
7.5   Vertical Line Charts   198
7.6   Horizontal Line Charts   205
7.7   Dot Plots   212
7.8   Combining the Plots   218

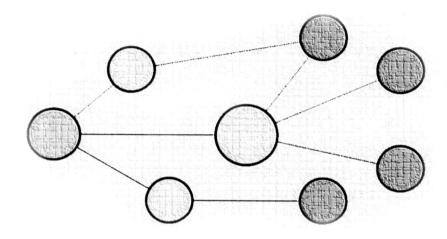

"There are no facts, only interpretations."
~ Friedrich Nietzsche

Chapter 7: Categorization Plots

In this chapter, we examine the categorization plots for SGPLOT and SGPANEL procedures. The unique feature of this class of plots is that the data is summarized by using a specified or default statistic. The basic syntax for each plot and the options will be covered in detail in this chapter. Plots within this group can be overlaid only with another plot in this group of the same orientation.

7.1 Introduction

The SGPLOT and SGPANEL procedures support the categorization plots listed below. Each of these plots support sum, mean, and frequency statistics. In addition, these plots support limit calculations for mean statistics. The limit statistics include the confidence limit of the mean (CLM), standard error (STDERR), and standard deviation (STDDEV). The confidence of the CLM statistic can be adjusted using the ALPHA option, and the number of STDERRs or STDDEVs can be specified using the NUMSTD option.

These plots cannot be used to draw pre-calculated limits. To draw custom limits on a bar chart, you should use a VBARPARM or HBARPARM plot (see Figure 4.11.1). For line charts, use a SERIES plot overlaid with a SCATTER plot (see Figure 4.6.10). For DOT plot, simply use a scatter plot (see Figure 4.4.7).

VBAR	HBAR	DOT
VLINE	HLINE	

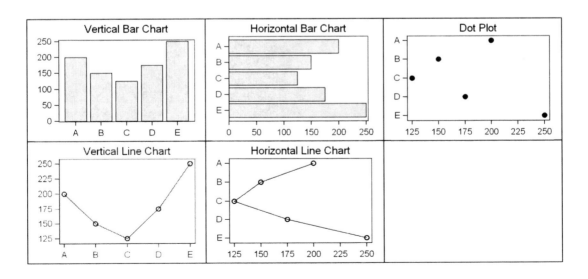

7.2 Categorization Plot Roles and Common Options

All of the categorization plot statements have the following syntax:

```
CategorizationPlotName required-data-roles </ options>;
```

All plot statements have required and optional parameters necessary to create the visual. Here is a list of common options for categorization plots:

Common Options:

ALPHA	=*value*	Confidence level of CLM statistic
ATTRID	=*string*	The attribute map ID used for the group variable
CATEGORYORDER	=*keyword*	RESPASC | RESPDESC
CLUSTERWIDTH	=*number*	Cluster width as a ratio of midpoint spacing
DATALABEL	<=*column*>	Display a scatter label for the Y variable or from an optional variable
DATALABELATTRS	=*text-attrs*	Appearance attributes for DataLabel
DISCRETEOFFSET	=*number*	Amount to offset all data primitives from the category midpoints
GROUPORDER	=*keyword*	DATA | ASCENDING | DESCENDING
LEGENDLABEL	=*string*	The label that appears in the legend to represent this (non-group) plot
LIMITATTRS	=*line-attrs*	Appearance attributes for the limits
LIMITS	=*keyword*	BOTH | UPPER | LOWER
LIMITSTAT	=*keyword*	CLM|STDDEV | STDERR
NAME	=*string*	Specifies a name for this statement. Other statements, such as a KEYLEGEND, can refer to a plot by its name.
NUMSTD	=*positive-integer*	Number of standard units to compute
STAT	=*keyword*	FREQ | MEAN | SUM
TRANSPARENCY	=*value*	Specifies the transparency for the visual elements
URL	=*string-column*	Contains URLs for drilldown
WEIGHT	=*num-column*	Used to weight observations

Common Boolean Options:

MISSING	Show missing category values
NOSTATLABEL	Suppress statistic on the axis label

7.3 Vertical Bar Charts

The VBAR plot shows the summarization of the response variable by category or the frequency count of the category. The syntax is:

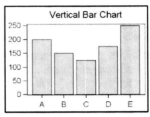
Vertical Bar Chart

```
vbar column </ options>;
```

By default, the data is summarized for each unique category value; however, user-defined formats can be used to create categorical ranges for summarization. One bar is drawn for each resulting category. If a group variable is specified, this bar can be subdivided by each group value, or the bar can be drawn as a cluster of smaller group bars (see the GROUPDISPLAY option).

Required Data Roles:
CATEGORY	column	Category variable

Optional Data Roles:
FREQ	=num-column	Frequency count for each observation
GROUP	=column	Group variable for each observation
RESPONSE	=num-column	Response values for the chart

Appearance Options:
FILLATTRS	=fill-attrs	Appearance attributes for Fill

Other Options:
BARWIDTH	=value	Width of bar a fraction of tick spacing
DATALABELPOS	=keyword	DATA | TOP | BOTTOM
DATASKIN	=keyword	Quasi-3-D effect for the bars (GLOSS, etc.)
GROUPDISPLAY	=keyword	STACK | CLUSTER

Boolean Options:
FILL | NOFILL	Controls bar fill
OUTLINE | NOOUTLINE	Controls bar outline

Standard common options are supported

Related Style Elements: The GraphDataDefault style element is used to draw the plot elements. When a GROUP variable is present, or when multiple plot statements are overlaid, the GraphData1-GraphData12 elements are used, one for each group or plot statement.

Figure 7.3.1: Bar Chart with Fill Attributes and Data Skins

This basic bar chart has been enhanced by specifying a particular color for the bar fill. In addition, the DATASKIN option is used to give the bars a modern, three-dimensional appearance while maintaining the visual accuracy of a two-dimensional bar chart. See the documentation for the different skins available.

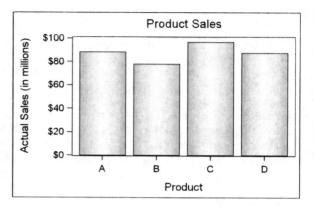

```
title "Product Sales";
proc sgplot data=sgbook.product_sales;
   vbar product / response=actual fillattrs=(color=grayee)
      dataskin=pressed;
   run;
```

Figure 7.3.2: Bar Chart with No Fill

This example uses the NOFILL option to create a "minimal ink" appearance that reproduces well in black and white publications.

There is also a NOOUTLINE option to remove the bar outlines when you only want the fill color.

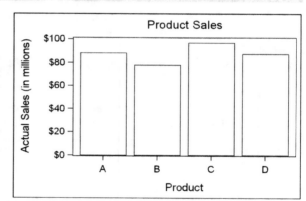

```
title "Product Sales";
proc sgplot data=sgbook.product_sales;
  vbar product / response=actual nofill;
  run;
```

Figure 7.3.3: Bar Chart with Data Labels

This example displays the bar response values above the bars by using the DATALABEL option. This option can be specified with a variable to display information other than the bar response values.

The DATALABELATTRS can be used to override the default text attributes of the labels.

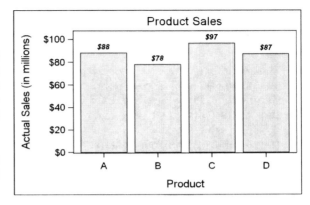

```
title "Product Sales";
proc sgplot data=sgbook.product_sales;
  vbar product / response=actual datalabel
      datalabelattrs=(weight=bold style=italic);
  run;
```

Figure 7.3.4: Bar Chart with Reference Line

Reference lines can be added to the bar chart by specifying a REFLINE statement. The order of the statements is significant.

If the REFLINE statement is specified before the VBAR statement, the line is drawn behind the bar chart. REFLINE statements specified after the VBAR statement are drawn in front of the bar chart as shown.

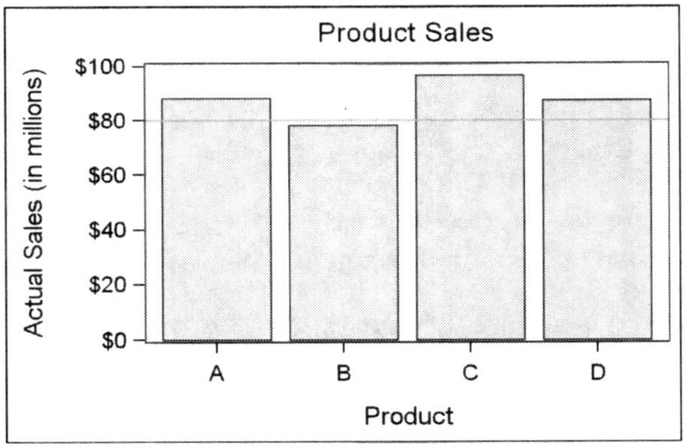

```
title "Product Sales";
proc sgplot data=sgbook.product_sales;
  vbar product / response=actual;
  refline 80;
  run;
```

Figure 7.3.5: Bar Chart with Confidence Limits

This example shows a bar chart showing the confidence limit of the mean (CLM). The standard error and standard deviation statistics are also available.

The ALPHA option can be used to specify the confidence for the CLM calculation. The default is 0.05. The default legend includes the limit statistic information.

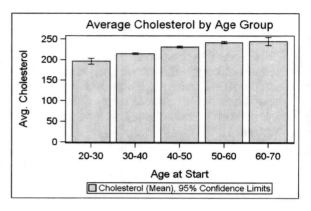

```
title "Average Cholesterol by Age Group";
proc sgplot data=sashelp.heart;
format AgeAtStart agefmt.;
  yaxis label="Avg. Cholesterol";
  vbar AgeAtStart / response=cholesterol stat=mean limitstat=clm
       alpha=0.05;
  run;
```

Figure 7.3.6: Bar Chart with an Upper Limit

You can use the LIMITS option to control which side of the limits are displayed.

This example uses the STDERR statistic with a NUMSTD multiplier to display two standard errors instead of one.

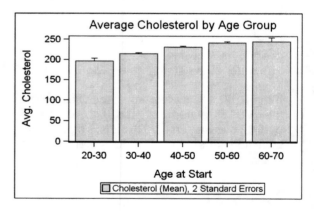

```
title "Average Cholesterol by Age Group";
proc sgplot data=sashelp.heart;
  format AgeAtStart agefmt.;
  yaxis label="Avg. Cholesterol";
  vbar AgeAtStart / response=cholesterol stat=mean limits=upper
       limitstat=stderr numstd=2;
  run;
```

Figure 7.3.7: Bar Chart with Limits and Label Positioning

This example shows the display of the bar and limits data.

The position of the data labels can be controlled by using the DATALABELPOS option.

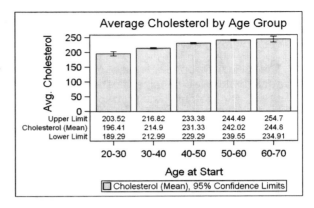

```
title "Average Cholesterol by Age Group";
proc sgplot data=sashelp.heart;
   format AgeAtStart agefmt.;
   yaxis label="Avg. Cholesterol";
   vbar AgeAtStart / response=cholesterol stat=mean limitstat=clm
        datalabel datalabelpos=bottom;
   run;
```

Figure 7.3.8: Bar Chart with Response Sorting

By default, the bars are sorted by the unformatted category value.

In this example, the CATEGORYORDER option is used to sort the bars by descending response value.

```
title "Product Sales";
proc sgplot data=sgbook.product_sales;
  vbar product / response=actual categoryorder=respdesc;
  run;
```

Figure 7.3.9: Bar Chart with Stacked Groups

The GROUP option is use to specify an additional level of classification in a bar chart. By default, the group values are represented as stacked bar segments. The data labels can be displayed above or below the chart.

The grouped bar attributes can be modified by using an attribute map (see Chapter 9) or by modifying an ODS style.

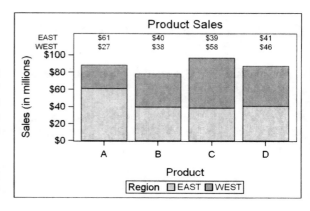

```
title "Product Sales";
proc sgplot data=sgbook.product_sales;
  yaxis label="Sales (in millions)";
  vbar product / response=actual group=region datalabel
      datalabelpos=top;
  run;
```

Figure 7.3.10: Bar Chart with Adjacent Groups

Instead of stacking the group values, the GROUPDISPLAY= CLUSTER option can be used to display the bars side by side.

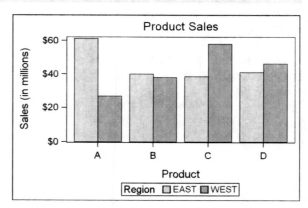

```
title "Product Sales";
proc sgplot data=sgbook.product_sales;
  yaxis label="Sales (in millions)";
  vbar product / response=actual group=region groupdisplay=cluster;
  run;
```

Figure 7.3.11: Bar Chart with Groups and Pattern Fills

The JOURNAL2 style is designed to be a black-and-white style. When this style is used with bar charts, fill patterns (SAS 9.3) are used in situations that need unique colors.

The MONOCHROMEPRINTER style is also a black-and-white style, but uses serif fonts instead of sans-serif fonts.

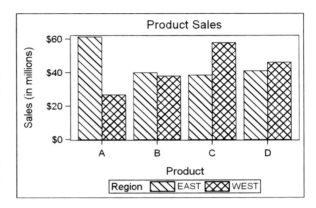

```
ods listing style=journal2;
title "Product Sales";
proc sgplot data=sgbook.product_sales;
   yaxis label="Sales (in millions)";
   vbar product / response=actual group=region groupdisplay=cluster;
   run;
```

Figure 7.3.12: Grouped Bar Chart using PROC SGPANEL

In the previous grouping examples, the group values are clustered around a category value. In this example, the category values are clustered around a group value.

The NOBORDER option is used to remove the border between cells. The LAYOUT =COLUMNLATTICE option displays all cells in one row.

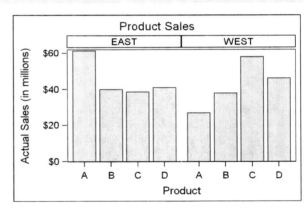

```
title "Product Sales";
proc sgpanel data=sgbook.product_sales;
   panelby region / layout=columnlattice noborder novarname onepanel;
   vbar product / response=actual;
   run;
```

Figure 7.3.13: Bullet Chart

This bullet chart is drawn using two overlaid bar charts. The first chart is drawn using the default bar width (0.85). The overlaid bar chart is drawn using a bar width of 0.3.

This example can be enhanced by adding transparency to the charts, enabling you to see the top of the first bars through the second bars.

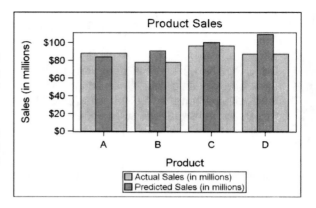

```
title "Product Sales";
proc sgplot data=sgbook.product_sales;
   yaxis label="Sales (in millions)";
   vbar product / response=actual;
   vbar product / response=predict barwidth=0.3;
   run;
```

Figure 7.3.14: Overlaid Bar Charts with Discrete Offsets

Overlaid bars can be placed side by side by using the DISCRETEOFFSET options. Bar widths have to be adjusted as shown.

A negative discrete offset moves the bars to the left; a positive value moves the bars to the right. The axis REVERSE option reverses this behavior.

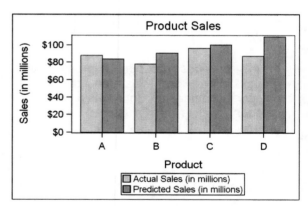

```
title "Product Sales";
proc sgplot data=sgbook.product_sales;
   yaxis label="Sales (in millions)";
   vbar product / response=actual barwidth=0.4 discreteoffset=-0.2;
   vbar product / response=predict barwidth=0.4 discreteoffset=0.2;
   run;
```

7.4 Horizontal Bar Charts

The HBAR plot shows the summarization of the response variable by category or the frequency count of the category. The syntax is:

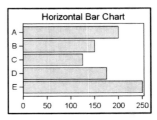

```
hbar column </ options>;
```

By default, the data is summarized for each unique category value; however, user-defined formats can be used to create categorical ranges for summarization. One bar is drawn for each resulting category. If a group variable is specified, this bar can be subdivided by each group value, or the bar can be drawn as a cluster of smaller group bars (see the GROUPDISPLAY option).

Required Data Roles:
CATEGORY	column	Category variable.

Optional Data Roles:
FREQ	=num-column	Frequency count for each observation
GROUP	=column	Group variable for each observation
RESPONSE	=num-column	Response values for the chart

Appearance Options:
FILLATTRS	=fill-attrs	Appearance attributes for fill

Other Options:
BARWIDTH	=value	Width of bar a fraction of tick spacing
DATASKIN	=keyword	Quasi-3D effect for the bars (GLOSS, etc.)
GROUPDISPLAY	=keyword	STACK	CLUSTER

Boolean Options:
FILL	NOFILL	Controls bar fill
OUTLINE	NOOUTLINE	Controls bar outline

Standard common options are supported

Related Style Elements: The GraphDataDefault style element is used to draw the plot elements. When a GROUP variable is present, or when multiple plot statements are overlaid, the GraphData1-GraphData12 elements are used, one for each group or plot statement.

Figure 7.4.1: Bar Chart with Fill Attributes and Data Skins

This basic bar chart has been enhanced by specifying a particular color for the bar fill. In addition, the DATASKIN option is used to give the bars a modern, three-dimensional appearance while maintaining the visual accuracy of a two-dimensional bar chart. See the documentation for the different skins available.

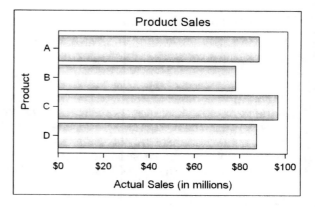

```
title "Product Sales";
proc sgplot data=sgbook.product_sales;
  hbar product / response=actual fillattrs=(color=grayee)
      dataskin=pressed;
  run;
```

Figure 7.4.2: Bar Chart with No Fill

This example uses the NOFILL option to create a "minimal ink" appearance that reproduces well in black and white publications.

There is also a NOOUTLINE option to remove the bar outlines when you only want the fill color.

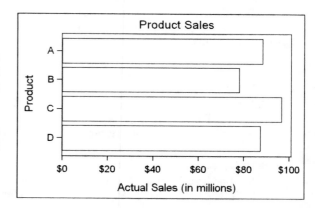

```
title "Product Sales";
proc sgplot data=sgbook.product_sales;
  hbar product / response=actual nofill;
  run;
```

Figure 7.4.3: Bar Chart with Data Labels

This example displays the bar response values above the bars by using the DATALABEL option. This can be specified with a variable to display information other than the bar response values.

The DATALABELATTRS is used to override the default text attributes of the labels. CATEGORYORDER is used to sort the bars by descending response.

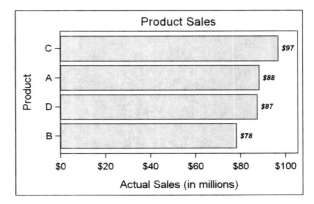

```
title "Product Sales";
proc sgplot data=sgbook.product_sales;
  hbar product / response=actual datalabel categoryorder=respdesc
      datalabelattrs=(weight=bold style=italic);
run;
```

Figure 7.4.4: Bar Chart with a Reference Line

Reference lines can be added to the bar chart by specifying a REFLINE statement. All plot statements are drawn in the order in which they are specified.

Since the REFLINE statement is specified after the HBAR statement, it is drawn in front of the bar chart.

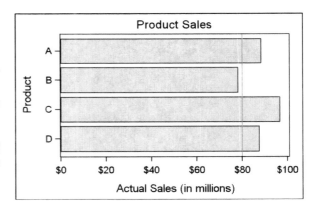

```
title "Product Sales";
proc sgplot data=sgbook.product_sales;
  hbar product / response=actual;
  refline 80 / axis=X;
run;
```

Figure 7.4.5: Bar Chart with Confidence Limits

This example shows a bar chart showing the confidence limit of the mean (CLM). The standard error and standard deviation statistics are also available.

The ALPHA option can be used to specify the confidence for the CLM calculation. The default is 0.05. The default legend includes the limit statistic information.

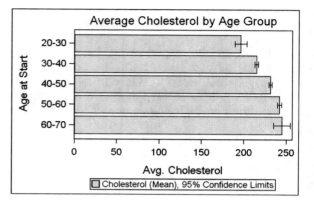

```
title "Average Cholesterol by Age Group";
proc sgplot data=sashelp.heart;
format AgeAtStart agefmt.;
  xaxis label="Avg. Cholesterol";
  hbar AgeAtStart / response=cholesterol stat=mean limitstat=clm
       alpha=0.05;
  run;
```

Figure 7.4.6: Bar Chart with Upper Limit

You can use the LIMITS option to control which side of the limits are displayed.

This example uses the STDERR statistic with a NUMSTD multiplier to display two standard errors instead of one.

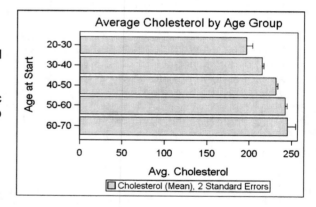

```
title "Average Cholesterol by Age Group";
proc sgplot data=sashelp.heart;
  format AgeAtStart agefmt.;
  xaxis label="Avg. Cholesterol";
  hbar AgeAtStart / response=cholesterol stat=mean limits=upper
       limitstat=stderr numstd=2;
  run;
```

Figure 7.4.7: Bar Chart with Groups

The GROUP option is use to specify an additional level of classification in a bar chart. By default, the group values are represented as stacked bar segments.

The bar segments are automatically assigned unique attributes for each group value. These attributes can be modified by using an attribute map (see Chapter 9) or by modifying an ODS style.

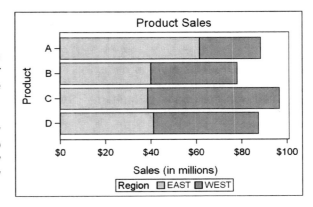

```
title "Product Sales";
proc sgplot data=sgbook.product_sales;
  xaxis label="Sales (in millions)";
  hbar product / response=actual group=region;
  run;
```

Figure 7.4.8: Bar Chart with Adjacent Groups

Instead of stacking the group values, the GROUPDISPLAY option can be used to CLUSTER the bars.

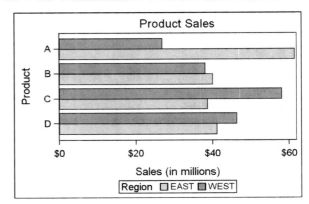

```
title "Product Sales";
proc sgplot data=sgbook.product_sales;
  xaxis label="Sales (in millions)";
  hbar product / response=actual group=region groupdisplay=cluster;
  run;
```

Figure 7.4.9: Bar Chart with Groups and Pattern Fills

The JOURNAL2 style is designed to be a black-and-white style. When this style is used with bar charts, fill patterns (SAS 9.3) are used in situations that need unique colors.

The MONOCHROMEPRINTER style is also a black-and-white style, but uses serif fonts instead of sans-serif fonts.

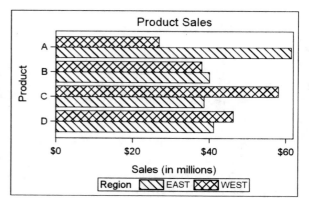

```
title "Product Sales";
proc sgplot data=sgbook.product_sales;
  xaxis label="Sales (in millions)";
  hbar product / response=actual group=region groupdisplay=cluster;
  run;
```

Figure 7.4.10: Grouped Bar Chart using PROC SGPANEL

In the previous grouping examples, the group values are clustered around a category value. In this example, the category values are clustered around a group value.

The NOBORDER option is used to remove the border between panel cells, and the LAYOUT=ROWLATTICE option ensures cells are in one column.

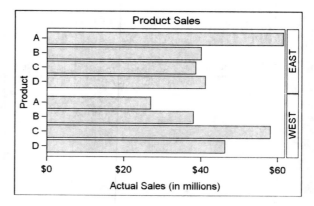

```
title "Product Sales";
proc sgpanel data=sgbook.product_sales;
  panelby region / layout=rowlattice noborder novarname onepanel;
  hbar product / response=actual;
  run;
```

Figure 7.4.11: Bullet Chart

This bullet chart is drawn using two overlaid bar charts. The first chart is drawn using the default bar width (0.85). The overlaid bar chart is drawn using a bar width of 0.3.

This example can be enhanced by adding transparency to the charts, enabling you to see the top of the first bars through the second bars.

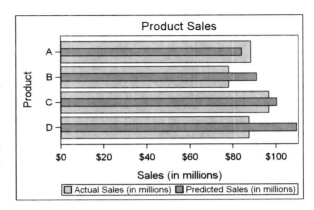

```
title "Product Sales";
proc sgplot data=sgbook.product_sales;
  xaxis label="Sales (in millions)";
  hbar product / response=actual;
  hbar product / response=predict barwidth=0.3;
run;
```

Figure 7.4.12: Overlaid Bar Charts with Discrete Offsets

Overlaid bars can be drawn side by side by using the DISCRETEOFFSET option. Bar widths have to be adjusted as shown.

A negative discrete offset moves the bars up; a positive value moves the bars down. The axis REVERSE option reverses this behavior. The bars can be overlapped.

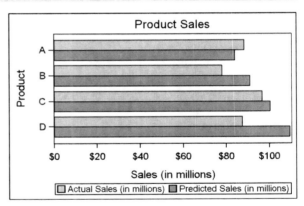

```
title "Product Sales";
proc sgplot data=sgbook.product_sales;
  xaxis label="Sales (in millions)";
  hbar product / response=actual barwidth=0.4 discreteoffset=-0.2;
  hbar product / response=predict barwidth=0.4 discreteoffset=0.2;
run;
```

7.5 Vertical Line Charts

The VLINE plot shows the summarization of the response variable by category or the frequency count of the category. The syntax is:

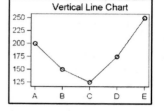

```
vline column </ options>;
```

By default, the data is summarized for each unique category value; however, user-defined formats can be used to create categorical ranges for summarization. One line is drawn connecting all resulting categories. If a group variable is specified, one line is drawn for each group value, either overlaid on the tick value or offset from the tick value to create a cluster effect (see the GROUPDISPLAY option).

Required Data Roles:

CATEGORY	column	Category variable

Optional Data Roles:

FREQ	=num-column	Frequency count for each observation
GROUP	=column	Group variable for each observation
RESPONSE	=column	Response values for the chart

Appearance Options:

CURVELABELATTRS	=text-attrs	Appearance attributes for the line labels
LINEATTRS	=line-attrs	Appearance attributes for the line
MARKERATTRS	=marker-attrs	Appearance attributes for the markers

Other Options:

CURVELABEL	<="string">	Label(s) for the line(s)
CURVELABELLOC	=keyword	INSIDE | OUTSIDE
CURVELABELPOS	=keyword	AUTO | MIN | MAX | START | END
DATALABELPOS	=keyword	DATA | TOP | BOTTOM
GROUPDISPLAY	=keyword	OVERLAY | CLUSTER

Boolean Options:

BREAK	Break the line at missing response values
MARKERS	Display markers on the data points
X2AXIS	Assign the plot to the secondary x-axis
Y2AXIS	Assign the plot to the secondary y-axis

Standard common options are supported

Related Style Elements: Same as VBAR and HBAR.

Figure 7.5.1: Basic Line Chart

This example shows a basic line chart using a MEAN statistic. The visual attributes of the line can be controlled using the LINEATTRS option.

The axis LABEL option is used to override the column label from the data set.

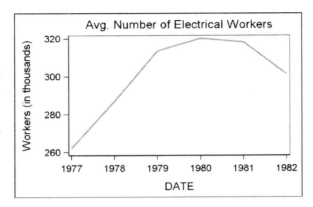

```
title "Avg. Number of Electrical Workers";
proc sgplot data=sashelp.workers;
   format date year.;
   yaxis label="Workers (in thousands)";
   vline date / response=electric stat=mean
         lineattrs=(color=gray thickness=2);
   run;
```

Figure 7.5.2: Line Chart with Markers

The MARKERS option displays markers on the data points along the line.

The MARKERATTRS option is used to control the attributes of the markers independently from the line attributes.

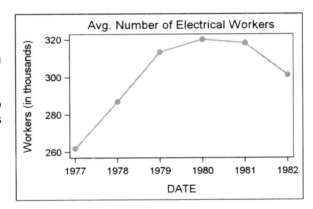

```
title "Avg. Number of Electrical Workers";
proc sgplot data=sashelp.workers;
   format date year.;
   yaxis label="Workers (in thousands)";
   vline date / response=electric stat=mean markers
         lineattrs=(color=gray thickness=2)
         markerattrs=(color=gray symbol=circlefilled);
   run;
```

Figure 7.5.3: Line Chart with Break

By default, line charts connect all of the non-missing response values. However, you might want to show the missing values by breaking the line using the BREAK option.

In this example, 1979 has all missing values. The line continues, starting at the next non-missing value.

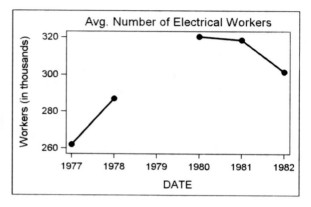

```
title "Avg. Number of Electrical Workers";
proc sgplot data=sgbook.workers_missing;
   format date year.;
   yaxis label="Workers (in thousands)";
   vline date / response=electric break stat=mean markers
         lineattrs=(thickness=2) markerattrs=(symbol=circlefilled);
run;
```

Figure 7.5.4: Line Chart with Data Labels

This example displays the line response values along the line by using the DATALABEL option. This option can be specified with a variable to display information other than the line response values.

The DATALABELATTRS can be used to override the default text attributes of the labels.

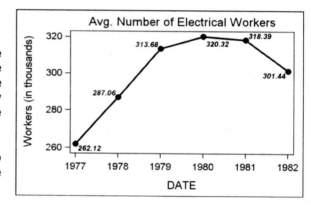

```
title "Avg. Number of Electrical Workers";
proc sgplot data=sashelp.workers;
   format date year.;
   yaxis label="Workers (in thousands)";
   vline date / response=electric stat=mean markers datalabel
         lineattrs=(thickness=2) markerattrs=(symbol=circlefilled)
         datalabelattrs=(weight=bold style=italic);
run;
```

Figure 7.5.5: Line Chart with a Reference Line

Reference lines can be added to the line chart by specifying a REFLINE statement. The order of the statements is significant.

Since the REFLINE statement is specified after the VLINE statement, the reference line is drawn in front of the line chart.

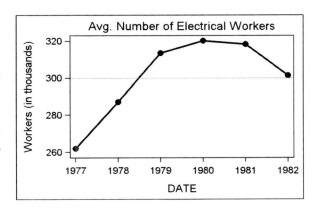

```
title "Avg. Number of Electrical Workers";
proc sgplot data=sashelp.workers;
  format date year.;
  yaxis label="Workers (in thousands)";
  vline date / response=electric stat=mean lineattrs=(thickness=2)
       markers markerattrs=(symbol=circlefilled);
  refline 300;
  run;
```

Figure 7.5.6: Line Chart with Confidence Limits

This example shows a line chart showing the confidence limit of the mean (CLM). The standard error and standard deviation statistics are also available.

The ALPHA option can be used to specify the confidence for the CLM calculation. The default is 0.05.

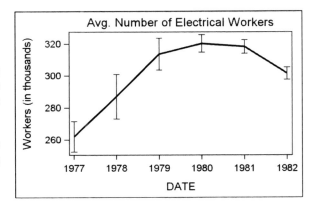

```
title "Avg. Number of Electrical Workers";
proc sgplot data=sashelp.workers noautolegend;
  format date year.;
  yaxis label="Workers (in thousands)";
  vline date / response=electric stat=mean limitstat=clm
            lineattrs=(thickness=2);
  run;
```

Figure 7.5.7: Line Chart with Upper Limits

You can use the LIMITS option to control which side of the limits are displayed.

This example uses the STDERR statistic with a NUMSTD multiplier to display two standard errors instead of one.

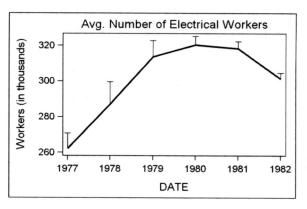

```
title "Avg. Number of Electrical Workers";
proc sgplot data=sashelp.workers noautolegend;
  format date year.;
  yaxis label="Workers (in thousands)";
  vline date / response=electric stat=mean limits=upper
          limitstat=stderr numstd=2 lineattrs=(thickness=2);
run;
```

Figure 7.5.8: Line Chart with CLM and Data Label Position

The DATALABELPOS option not only moves the data labels above or below the plot area, it will also display the limits (if applicable).

The LABEL statement is used to control the default column label for ELECTRIC (needed for the data label table), but this label is further overridden by the LABEL option on the YAXIS statement.

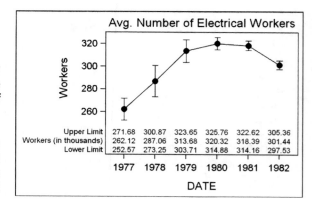

```
title "Avg. Number of Electrical Workers";
proc sgplot data=sashelp.workers;
  format date year.;
  label electric="Workers (in thousands)";
  yaxis label="Workers";
  vline date / response=electric markers stat=mean limitstat=clm
     markerattrs=(symbol=circlefilled) datalabel datalabelpos=bottom;
run;
```

Figure 7.5.9: Grouped Line Chart

The GROUP option is use to specify an additional level of classification in a line chart. The plot lines and markers are automatically assigned unique attributes for each group value. These attributes can be modified by using an attribute map (see Chapter 9) or by modifying an ODS style. Data labels can be displayed above or below the chart, as well as along the line.

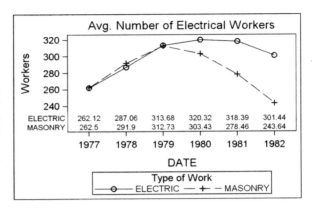

```
title "Avg. Number of Electrical Workers";
proc sgplot data=sgbook.workers_group; /* Transpose of sashelp.workers */
  label _name_="Type of Work";
  format date year.;
  yaxis label="Workers";
  vline date / response=col1 markers stat=mean
        group=_name_ datalabel datalabelpos=bottom;
  run;
```

Figure 7.5.10: Line Chart with Curve Labels

In this grouped plot example, the CURVELABEL option uses the group values to label the lines. When a line chart has no group variable, you can specify a curve label by specifying CURVELABEL="<your string>".

Curve labels can be positioned inside or outside the data area and at the beginning or end of the plot line.

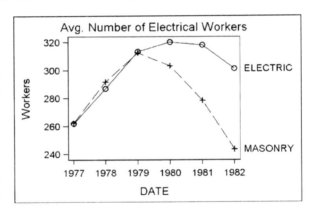

```
title "Avg. Number of Electrical Workers";
proc sgplot data=sgbook.workers_group;
  format date year.;
  yaxis label="Workers";
  vline date / response=col1 markers stat=mean
        group=_name_ curvelabel curvelabelloc=outside;
  run;
```

Figure 7.5.11: Overlaid Line Charts

This example plots two response variables by overlaying a line chart for each variable.

Each statement has its own set of visual attributes and can be assigned to different axes. Also, the legend label can be changed for each chart by using the LEGENDLABEL option.

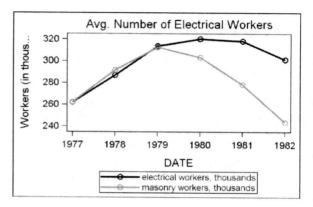

```
title "Avg. Number of Electrical Workers";
proc sgplot data=sashelp.workers;
  format date year.;
  yaxis label="Workers (in thousands)";
  vline date / response=electric stat=mean markers markerattrs=(color=black
      symbol=circle) lineattrs=(color=black thickness=2);
  vline date / response=masonry stat=mean markers markerattrs=(color=gray
      symbol=circle) lineattrs=(color=gray pattern=solid thickness=2);
run;
```

Figure 7.5.12: Overlaid Line Charts with Discrete Offset

Normally, response values of overlaid lines are drawn at the tick positions along the category axis, which can cause overlap. To move the values off of the tick positions, use the DISCRETEOFFSET option.

Negative values down to -0.5 move the points to the left. Positive values up to 0.5 move the points to the right.

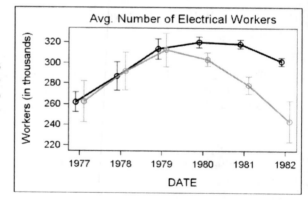

```
title "Avg. Number of Electrical Workers";
proc sgplot data=sashelp.workers noautolegend;
  format date year.;
  yaxis label="Workers (in thousands)"; xaxis offsetmin=0.05 offsetmax=0.05;
  vline date / response=electric stat=mean markers limitstat=clm discreteoffset=-0.1
      markerattrs=(color=black symbol=circle) lineattrs=(color=black thickness=2);
  vline date / response=masonry stat=mean markers discreteoffset=0.1 limitstat=clm
      markerattrs=(color=gray symbol=circle) lineattrs=(color=gray pattern=solid
      thickness=2);
run;
```

7.6 Horizontal Line Charts

The HLINE plot shows the summarization of the response variable by category or the frequency count of the category. The syntax is:

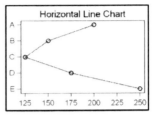

```
hline column </ options>;
```

By default, the data is summarized for each unique category value; however, user-defined formats can be used to create categorical ranges for summarization. One line is drawn connecting all resulting categories. If a group variable is specified, one line is drawn for each group value, either overlaid on the tick value or offset from the tick value to create a cluster effect (see the GROUPDISPLAY option).

Required Data Roles:
    CATEGORY   *column*   Category variable

Optional Data Roles:
    FREQ   *=num-column*   Frequency count for each observation
    GROUP   *=column*   Group variable for each observation
    RESPONSE   *=column*   Response values for the chart

Appearance Options:
    CURVELABELATTRS   *=text-attrs*   Appearance attributes for the line labels
    LINEATTRS   *=line-attrs*   Appearance attributes for the line
    MARKERATTRS   *=marker-attrs*   Appearance attributes for the markers

Other Options:
    CURVELABEL   <="string"\>   Label(s) for the line(s)
    CURVELABELLOC   *=keyword*   INSIDE | OUTSIDE
    CURVELABELPOS   *=keyword*   AUTO | MIN | MAX | START | END
    GROUPDISPLAY   *=keyword*   OVERLAY | CLUSTER

Boolean Options:
    BREAK   Break the line at missing response values
    MARKERS   Display markers on the data points
    X2AXIS   Assign the plot to the secondary x-axis
    Y2AXIS   Assign the plot to the secondary y-axis

Standard common options are supported

Related Style Elements: Same as VBAR and HBAR.

Figure 7.6.1: Basic Line Chart

This example shows a basic line chart using a MEAN statistic. The visual attributes of the line can be controlled using the LINEATTRS option.

The axis LABEL option is used to override the column label from the data set.

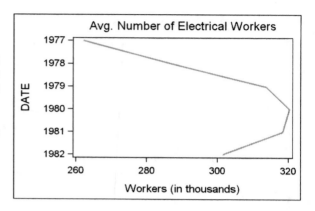

```
title "Avg. Number of Electrical Workers";
proc sgplot data=sashelp.workers;
  format date year.;
  yaxis grid;
  xaxis label="Workers (in thousands)";
  hline date / response=electric stat=mean
        lineattrs=(color=gray thickness=2);
  run;
```

Figure 7.6.2: Line Chart with Markers

The MARKERS option displays markers on the data points along the line.

The MARKERATTRS option is used to control the attributes of the markers independently from the line attributes.

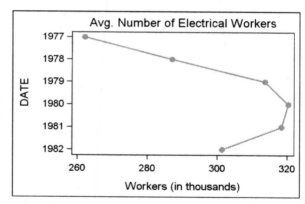

```
title "Avg. Number of Electrical Workers";
proc sgplot data=sashelp.workers;
format date year.;
  yaxis grid;
  xaxis label="Workers (in thousands)";
  hline date / response=electric stat=mean lineattrs=(color=gray thickness=2)
        markerattrs=(color=gray symbol=circlefilled) markers;
  run;
```

Figure 7.6.3: Line Chart with Break

By default, line charts connect all of the non-missing response values. However, you might want to show the missing values by breaking the line using the BREAK option.

In this example, 1979 has all missing values. The line continues, starting at the next non-missing value.

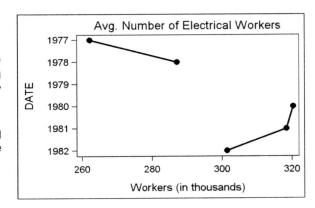

```
title "Avg. Number of Electrical Workers";
proc sgplot data=sgbook.workers_missing;
   format date year.;
   yaxis grid;
   xaxis label="Workers (in thousands)";
   hline date / response=electric break stat=mean lineattrs=(thickness=2)
      markers markerattrs=(symbol=circlefilled);
   run;
```

Figure 7.6.4: Line Chart with Data Labels

This example displays the line response values along the line by using the DATALABEL option. This option can be specified with a variable to display information other than the line response values.

The DATALABELATTRS can be used to override the default text attributes of the labels.

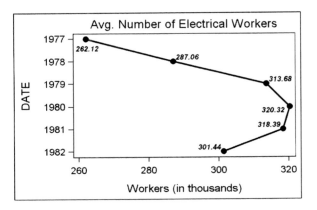

```
title "Avg. Number of Electrical Workers";
proc sgplot data=sashelp.workers;
   format date year.;
   yaxis grid;
   xaxis label="Workers (in thousands)";
   hline date / response=electric stat=mean markers lineattrs=(thickness=2)
      markerattrs=(symbol=circlefilled) datalabel
      datalabelattrs=(weight=bold style=italic);
   run;
```

Figure 7.6.5: Line Chart with a Reference Line

Reference lines can be added to the line chart by specifying a REFLINE statement. The order of the statements is significant.

Since the REFLINE statement is specified after the HLINE statement, the reference line is drawn in front of the line chart.

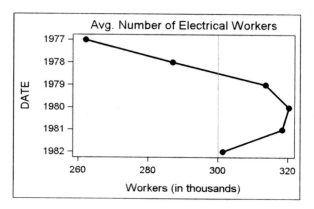

```
title "Avg. Number of Electrical Workers";
proc sgplot data=sashelp.workers;
   format date year.;
   xaxis label="Workers (in thousands)";
   hline date / response=electric stat=mean lineattrs=(thickness=2) markers
        markerattrs=(symbol=circlefilled);
   refline 300 / axis=X;
run;
```

Figure 7.6.6: Line Chart with Confidence Limits

This example shows a line chart showing the confidence limit of the mean (CLM). The standard error and standard deviation statistics are also available.

The ALPHA option can be used to specify the confidence for the CLM calculation. The default is 0.05.

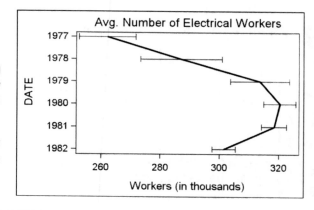

```
title "Avg. Number of Electrical Workers";
proc sgplot data=sashelp.workers noautolegend;
   format date year.;
   xaxis label="Workers (in thousands)" grid;
   hline date / response=electric stat=mean limitstat=clm
            lineattrs=(thickness=2);
run;
```

Figure 7.6.7: Line Chart with Upper Limits

You can use the LIMITS option to control which side of the limits are displayed.

This example uses the STDERR statistic with a NUMSTD multiplier to display two standard errors instead of one.

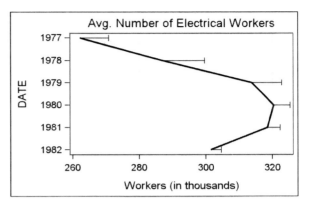

```
title "Avg. Number of Electrical Workers";
proc sgplot data=sashelp.workers noautolegend;
   format date year.;
   xaxis label="Workers (in thousands)" grid;
   hline date / response=electric stat=mean limits=upper
                limitstat=stderr numstd=2 lineattrs=(thickness=2);
   run;
```

Figure 7.6.8: Line Chart with Response Sorting

Use the CATEGORYORDER option to order your category values based on ascending or descending response values.

In this example, the products with the highest sales are sorted to the top of the axis.

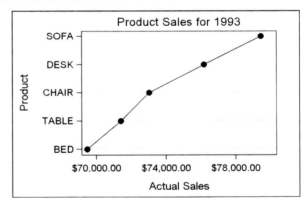

```
title "Product Sales for 1993";
proc sgplot data=sashelp.prdsale (where=(year=1993));
   yaxis grid;
   hline product / response=actual categoryorder=respdesc markers
        markerattrs=(symbol=circlefilled);
   run;
```

Figure 7.6.9: Grouped Line Chart

The GROUP option is used to specify an additional level of classification in a line chart. The plot lines and markers are automatically assigned unique attributes for each group value.

These attributes can be modified by using an attribute map (see Chapter 9) or by modifying an ODS style.

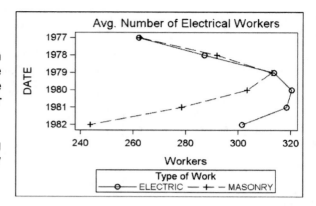

```
title "Avg. Number of Electrical Workers";
proc sgplot data=sgbook.workers_group;
   format date year.;
   label _name_="Type of Work";
   xaxis label="Workers";
   hline date / response=col1 markers group=_name_;
   run;
```

Figure 7.6.10: Line Chart with Curve Labels

This grouped plot example uses the CURVELABEL option to label the lines.

When a line chart has no group variable, you can specify a curve label by specifying CURVELABEL="<string>". Curve labels can be positioned inside or outside the data area and at the beginning or end of the plot line.

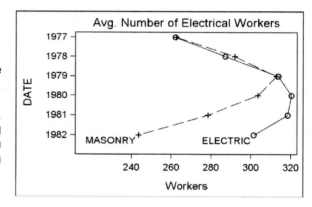

```
title "Avg. Number of Electrical Workers";
proc sgplot data=sgbook.workers_group;
   format date year.;
   xaxis label="Workers";
   hline date / response=col1 markers stat=mean
         group=_name_ curvelabel;
   run;
```

Figure 7.6.11: Overlaid Line Charts

This example plots two response variables by overlaying a line chart for each variable.

Each statement has its own set of visual attributes, and can be assigned to different axes. Also, the legend label can be changed for each chart by using the LEGENDLABEL option.

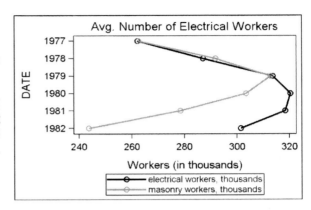

```
title "Avg. Number of Electrical Workers";
proc sgplot data=sashelp.workers;
   format date year.;
   xaxis label="Workers (in thousands)";
   hline date / response=electric stat=mean markers markerattrs=(color=black
         symbol=circle) lineattrs=(color=black thickness=2);
   hline date / response=masonry stat=mean markers markerattrs=(color=gray
         symbol=circle) lineattrs=(color=gray pattern=solid thickness=2);
   run;
```

Figure 7.6.12: Overlaid Line Charts with Discrete Offset

Normally, response values of overlaid lines are drawn at the tick positions along the category axis, which can cause overlap.

To move the values off of the tick positions, use the DISCRETEOFFSET option. Negative values down to -0.5 move the points down. Positive values up to 0.5 move the points up.

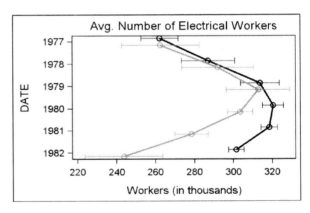

```
title "Avg. Number of Electrical Workers";
proc sgplot data=sashelp.workers;
   format date year.;
   xaxis label="Workers (in thousands)"; xaxis offsetmin=0.05 offsetmax=0.05;
   hline date / response=electric stat=mean markers limitstat=clm discreteoffset=-0.1
         markerattrs=(color=black symbol=circle) lineattrs=(color=black thickness=2);
   hline date / response=masonry stat=mean markers discreteoffset=0.1 limitstat=clm
         markerattrs=(color=gray symbol=circle) lineattrs=(color=gray pattern=solid
         thickness=2);
   run;
```

7.7 Dot Plots

The dot plot shows the summarization of the response variable by category or the frequency count of the category. The syntax is:

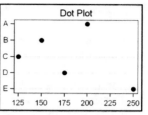

```
dot column </ options>;
```

By default, the data is summarized for each unique category value; however, user-defined formats can be used to create categorical ranges for summarization. One marker is drawn for resulting category. If a group variable is specified, one marker is drawn for each group value, either overlaid on the tick value or offset from the tick value to create a cluster effect (see the GROUPDISPLAY option).

Required Data Roles:
CATEGORY	*column*	Category variable

Optional Data Roles:
FREQ	=*num-column*	Frequency count for each observation
GROUP	=*column*	Group variable for each observation
RESPONSE	=*column*	Response values for the chart

Appearance Options:
MARKERATTRS	=*marker-attrs*	Appearance attributes for the markers

Other Options:
GROUPDISPLAY	=*keyword*	OVERLAY	CLUSTER

Boolean Options:
X2AXIS	Assign the plot to the secondary x-axis
Y2AXIS	Assign the plot to the secondary y-axis

Standard common options are supported

Related Style Elements: Same as VBAR and HBAR.

Figure 7.7.1: Basic Dot Plot

Dot plots are highly effective plots for comparing a large number of category values.

This example shows a basic dot plot using a MEAN statistic. The axis LABEL option is used to override the column label from the data set.

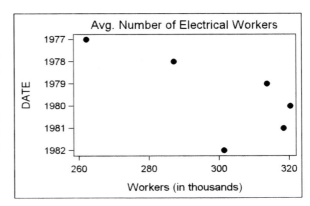

```
title "Avg. Number of Electrical Workers";
proc sgplot data=sashelp.workers;
   format date year.;
   xaxis label="Workers (in thousands)";
   dot date / response=electric stat=mean;
run;
```

Figure 7.7.2: Dot Plot with Marker Attributes

The MARKERATTRS option is used to control the attributes of the markers in the plot.

In this example, the marker shape and the marker size are modified.

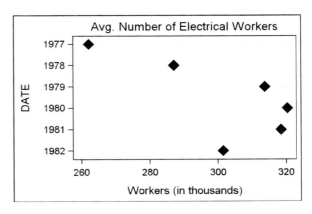

```
title "Avg. Number of Electrical Workers";
proc sgplot data=sashelp.workers;
   format date year.;
   xaxis label="Workers (in thousands)";
   dot date / response=electric stat=mean
      markerattrs=(symbol=diamondfilled size=11px);
run;
```

Figure 7.7.3: Dot Plot with Data Labels

This example displays the plot response values by using the DATALABEL option. This option can be specified with a variable to display information other than the plot response values.

The DATALABELATTRS can be used to override the default text attributes of the labels.

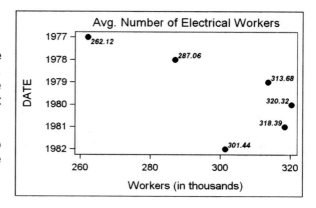

```
title "Avg. Number of Electrical Workers";
proc sgplot data=sgbook.workers_missing;
  format date year.;
  xaxis label="Workers (in thousands)";
  dot date / response=electric stat=mean datalabel
    datalabelattrs=(weight=bold style=italic);
run;
```

Figure 7.7.4: Dot Plot with a Reference Line

Reference lines can be added to the dot plot by specifying a REFLINE statement.

The order of the statements is significant. If the REFLINE statement is specified before the DOT statement, the line is drawn behind the dot plot. REFLINE statements specified after the DOT statement are drawn in front of the dot plot.

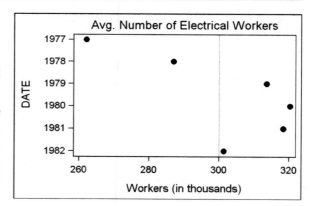

```
title "Avg. Number of Electrical Workers";
proc sgplot data=sashelp.workers;
  format date year.;
  xaxis label="Workers (in thousands)";
  dot date / response=electric stat=mean;
  refline 300 / axis=X;
run;
```

Figure 7.7.5: Dot Plot with Confidence Limits

This example shows a dot plot showing the confidence limit of the mean (CLM). The standard error and standard deviation statistics are also available.

The ALPHA option can be used to specify the confidence for the CLM calculation. The default is 0.05.

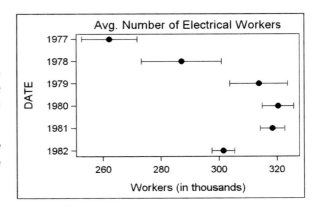

```
title "Avg. Number of Electrical Workers";
proc sgplot data=sashelp.workers noautolegend;
   format date year.;
   xaxis label="Workers (in thousands)" grid;
   dot date / response=electric stat=mean limitstat=clm;
run;
```

Figure 7.7.6: Dot Plot with Upper Limits

You can use the LIMITS option to control which side of the limits are displayed.

This example uses the STDERR statistic with a NUMSTD multiplier to display two standard errors instead of one.

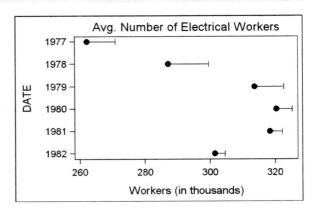

```
title "Avg. Number of Electrical Workers";
proc sgplot data=sashelp.workers noautolegend;
   format date year.;
   xaxis label="Workers (in thousands)" grid;
   dot date / response=electric stat=mean limits=upper
              limitstat=stderr numstd=2;
run;
```

Figure 7.7.7: Dot Plot with Response Sorting

Use the CATEGORYORDER option to order your category values based on ascending or descending response values.

In this example, the products with the highest sales are sorted to the top of the axis.

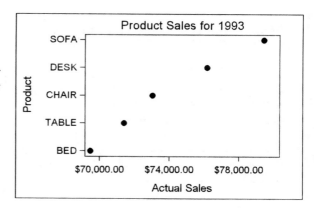

```
title "Product Sales for 1993";
proc sgplot data=sashelp.prdsale (where=(year=1993));
   dot product / response=actual categoryorder=respdesc;
   run;
```

Figure 7.7.8: Grouped Dot Plot

The GROUP option is used to specify an additional level of classification in a dot plot. The plot markers are automatically assigned unique attributes for each group value.

These marker attributes can be modified by using an attribute map (see Chapter 9) or by modifying an ODS style.

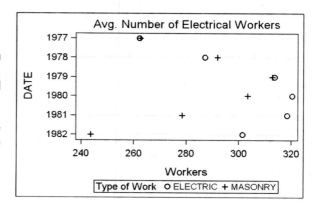

```
title "Avg. Number of Electrical Workers";
proc sgplot data=sgbook.workers_group;
   format date year.;
   label _name_="Type of Work";
   xaxis label="Workers";
   dot date / response=col1 group=_name_ stat=mean;
   run;
```

Figure 7.7.9: Overlaid Dot Plots

This example plots two response variables by overlaying a dot plot for each variable. Each statement has its own set visual attributes and can be assigned to different axes.

The legend label can be changed for each plot by using the LEGENDLABEL option.

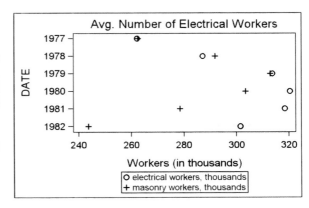

```
title "Avg. Number of Electrical Workers";
proc sgplot data=sashelp.workers;
   format date year.;
   xaxis label="Workers (in thousands)";
   dot date / response=electric stat=mean;
   dot date / response=masonry stat=mean;
   run;
```

Figure 7.7.10: Dot Plot with Discrete Offsets

Normally, response values of overlaid dot plots are drawn at the tick positions along the category axis, which can cause overlap.

To move the values off of the tick positions, use the DISCRETEOFFSET option. Negative values down to -0.5 move the points down. Positive values up to 0.5 move them up.

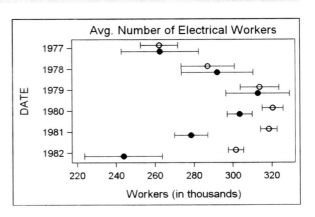

```
title "Avg. Number of Electrical Workers";
proc sgplot data=sashelp.workers;
   format date year.;
   xaxis label="Workers (in thousands)";
   yaxis offsetmin=0.075 offsetmax=0.075;
   dot date / response=electric stat=mean limitstat=clm
           discreteoffset=-0.15 markerattrs=(symbol=circle);
   dot date / response=masonry stat=mean discreteoffset=0.15
           limitstat=clm markerattrs=(symbol=circlefilled);
   run;
```

7.8 Combining the Plots

Figure 7.8.1: Bar-Line Overlay

In this example, a line chart is overlaid on a bar chart. The line chart is assigned to the Y2 axis to display the weight independently from the cholesterol range.

The data labels are displayed for each chart and positioned above and below the chart. The Y2 axis minimum is explicitly set to zero to synchronize with the y-axis range.

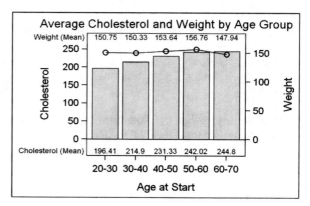

```
title "Average Cholesterol and Weight by Age Group";
proc sgplot data=sashelp.heart noautolegend;
  format AgeAtStart agefmt.;
  yaxis label="Cholesterol";
  y2axis label="Weight" min=0;
  vbar AgeAtStart / response=cholesterol stat=mean datalabel
       datalabelpos=bottom;
  vline AgeAtStart / response=weight stat=mean y2axis markers
       datalabel datalabelpos=top;
run;
```

Figure 7.8.2: HLINE-Dot Overlay

In this example, the HLINE draws the actual values in descending order. The line creates a visual separation for evaluating the predicted values.

Predicted values to the left of the line mean that sales exceeded the predicted value. Predicted values to the right of the line mean that sales fell short of the predicted value.

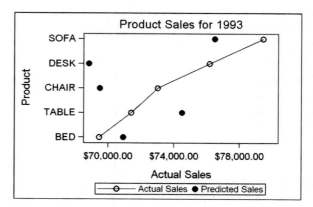

```
title "Product Sales for 1993";
proc sgplot data=sashelp.prdsale (where=(year=1993));
  hline product / response=actual categoryorder=respdesc markers;
  dot product / response=predict;
run;
```

Chapter 8

Axes, Legends, and Insets

8.1 Introduction 221
8.2 Linear Axis 223
8.3 Log Axis 225
8.4 Time Axis 227
8.5 Discrete Axis 229
8.6 Legends 230
8.7 Insets 232

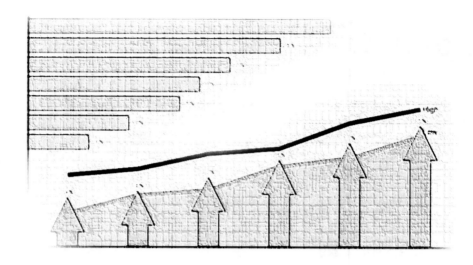

*"Never interrupt your enemy
when he is making a mistake."
~ Napoleon Bonaparte*

Chapter 8: Axes, Legends, and Insets

In this chapter, we examine the options for controlling the axes, legends, and insets. You have seen many of these options used throughout the examples in this book. This chapter will focus on the capabilities you have not yet seen.

8.1 Introduction

The SGPLOT procedure supports up to four axes (XAXIS, YAXIS, X2AXIS, and Y2AXIS), while the SGPANEL procedure supports only two (ROWAXIS and COLAXIS). The axis statements are contained within the procedure code and can be specified in any order. For the secondary axes to appear in SGPLOT, a plot must be assigned to them using the X2AXIS or Y2AXIS plot options.

The SGPLOT and SGPANEL procedures support four types of axes: LINEAR, LOG, TIME, and DISCRETE. The axis type is chosen automatically based on the chart type, the data, and the formats used; however, the LOG type is never chosen automatically. If you override the default axis type, be sure you do not specify an axis type that is incompatible with your chart type; otherwise, your chart will not draw. For example, if you request a vertical bar chart, the x-axis type is automatically set to DISCRETE. If you attempt to set the x-axis type to any other value, you will get a note in the log and the chart will not draw.

The following axis options can be used regardless of axis type. Axis options that apply to a specific type will be addressed in the following sections.

Common Axis Options:

DISPLAY	=display-value	The features of the axis to display
GRID		Control the display of grid lines
LABEL	=string	The label of the axis
LABELATTRS	=text-attrs	Text attributes for the axis label
OFFSETMIN	=number	Tick offset preceding the lowest value on the axis
OFFSETMAX	=number	Tick offset following the highest value on the axis
REFTICKS		Mirrors the ticks on the opposite axis
REVERSE		Reverses the order of the tick marks
TYPE	=keyword	LINEAR | LOG | TIME | DISCRETE
VALUEATTRS	=text-attrs	Text attributes for the axis tick values

The KEYLEGEND statement is used to specify the content and the attributes of custom legends, as well as control the attributes of automatic legends. This statement works the same way in both the SGPLOT and SGPANEL procedures, with the exception of the LOCATION option. The LOCATION option is not available in the SGPANEL procedure because internal legends are not supported. The SGSCATTER procedure has a LEGEND option that supports a subset of the options in the following legend table. See the documentation for more details.

If the legend is located inside, and the POSITION option is not specified, the legend will automatically position itself in an area with the least amount of data collision. This feature is very useful for batch programs. If the plot data varies significantly from one run to the next, the legend will move to the best possible position without any program change.

Legend Options:

ACROSS	=integer	Specifies the number of columns in the legend
BORDER / NOBORDER		Controls the display of the legend border
DOWN	=integer	Specifies the number of rows in the legend
LOCATION	=keyword	INSIDE | OUTSIDE
POSITION	=keyword	Specifies the legend position within the legend location (TOP, BOTTOM, etc.)
TITLE	=string	Specifies a title for the legend
TITLEATTRS	=text-attrs	Specifies text attributes for the legend title
VALUEATTRS	=text-attrs	Specifies text attributes for the legend values

The INSET statement is used to add additional text or small tables to an SGPLOT graph. As the statement name suggests, the inset is always drawn within the plot area of the graph. The information in the inset is specified directly on the statement. See the inset examples for more details.

If the POSITION option is not specified, the inset will automatically position itself in an area with the least amount of data collision. This feature is very useful for batch programs. If the plot data varies significantly from one run to the next, the inset will move to the best possible position without any program change.

Inset Options:

BORDER / NOBORDER		Controls the display of the inset border
LABELALIGN	=keyword	LEFT | RIGHT | CENTER
POSITION	=keyword	Specifies the inset position within the data area (TOP, BOTTOM, etc.)
TEXTATTRS	=text-attrs	Text attributes for the inset text
TITLE	= string	Specifies a title for the inset
TITLEATTRS	=text-attrs	Text attributes for the inset title
VALUEALIGN	=keyword	LEFT | RIGHT | CENTER

8.2 Linear Axis

Figure 8.2.1: Specifying Axis Min/Max

In this example, the MIN and MAX options are used to set the data range of the x-axis. These options set the extent of the data range but not necessarily the first and last tick values.

Appropriate tick values are automatically calculated to fit within the requested range. These options are available on all axis types except DISCRETE.

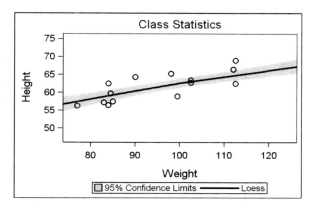

```
title "Class Statistics";
proc sgplot data=sashelp.class;
   xaxis min=75 max=125;
   loess x=weight y=height / clm;
   run;
```

Figure 8.2.2: Specifying a Tick List

In this example, the VALUES option is used to set both the extents of the axis and the tick values to display. The tick values can be specified as a list, a sequence, or both.

If the values do not have equal intervals, the values will still be placed in the correct numerical location along the axis. This option is supported by LINEAR and TIME axes.

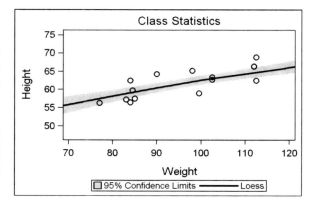

```
title "Class Statistics";
proc sgplot data=sashelp.class;
   xaxis values=(70 to 120 by 10);
   loess x=weight y=height / clm;
   run;
```

Figure 8.2.3: The VALUESHINT Option

By default, the low and high values on the VALUES option set the extents of the axis.

The VALUESHINT option can be used to show the full data extent is displayed, and the tick values in the list are drawn on the axis if they fall within the data extent. This option is supported on LINEAR and TIME axes.

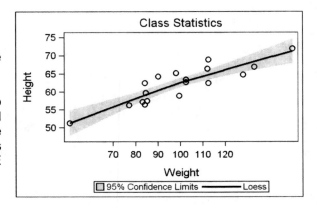

```
title "Class Statistics";
proc sgplot data=sashelp.class;
  xaxis values=(70 to 120 by 10) valueshint;
  loess x=weight y=height / clm;
run;
```

Figure 8.2.4: The INTEGER Option

If an axis displays a small data range, it is possible that tick values with decimal values can appear.

To prevent this, you can use the INTEGER option to force the axis to display only integer values.

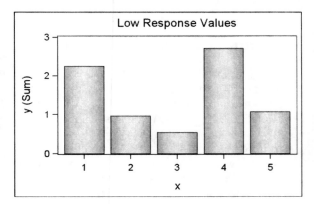

```
title "Low Response Values";
proc sgplot data=sgbook.random;
  yaxis integer;
  vbar x / response=y dataskin=pressed;
run;
```

8.3 Log Axis

Figure 8.3.1: Log Expand Style

In this example, the log axis shows the expanded form of the log exponent.

The MINOR option is specified to display the minor tick marks on the axis.

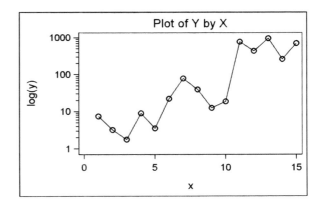

```
title "Plot of Y by X";
proc sgplot data=sgbook.logdata;
  xaxis integer;
  yaxis type=log logstyle=LogExpand label="log(y)" minor;
  series x=x y=y / markers;
  run;
```

Figure 8.3.2: Log Exponent Style

In this example, the log axis displays the log exponent instead of the expanded form.

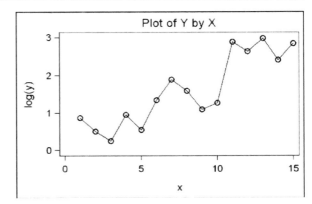

```
title "Plot of Y by X";
proc sgplot data=sgbook.logdata;
  xaxis integer;
  yaxis type=log logstyle=LogExponent label="log(y)";
  series x=x y=y / markers;
  run;
```

Figure 8.3.3: Log Linear Style

With the LINEAR style of log axis, the major tick marks are placed at non-uniform locations covering the range of the axis.

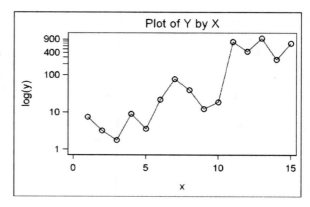

```
title "Plot of Y by X";
proc sgplot data=sgbook.logdata;
  xaxis integer;
  yaxis type=log logstyle=Linear label="log(y)";
  series x=x y=y / markers;
  run;
```

Figure 8.3.4: The LOGBASE Option

This option sets the base for the log calculations. The three valid values are 10, 2, and e.

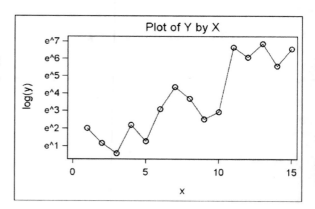

```
title "Plot of Y by X";
proc sgplot data=sgbook.logdata;
  xaxis integer;
  yaxis type=log logstyle=LogExpand label="log(y)" logbase=e;
  series x=x y=y / markers;
  run;
```

8.4 Time Axis

Figure 8.4.1: The INTERVAL Option

By default, the time axis calculates the best time interval and format. You can use the INTERVAL option to override the default interval.

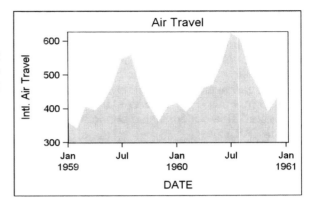

```
title "Air Travel";
proc sgplot data=sashelp.air;
  where date >= '01jan1959'd and date <='01jan1961'd;
  xaxis offsetmin=0 interval=semiyear;
  yaxis offsetmin=0 label="Intl. Air Travel";
  band x=date upper=air lower=300;
  run;
```

Figure 8.4.2: The NOTIMESPLIT Option

By default, the horizontal time axis extracts the highest time order and displays it on a second line to save space.

The NOTIMESPLIT option prevents the extraction. Compare this example with Figure 8.4.1. In this example, the year is kept with each tick value.

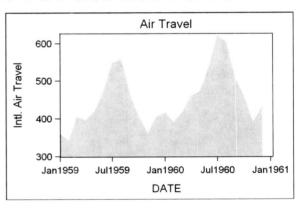

```
title "Air Travel";
proc sgplot data=sashelp.air;
  where date >= '01jan1959'd and date <='01jan1961'd;
  xaxis offsetmin=0 interval=semiyear notimesplit;
  yaxis offsetmin=0 label="Intl. Air Travel";
  band x=date upper=air lower=300;
  run;
```

Figure 8.4.3: The TICKVALUEFORMAT Option

By default, the time axis determines the best format to use based on the interval.

You can override this behavior by specifying a format using the TICKVALUEFORMAT option. This option is also supported by the LINEAR axis.

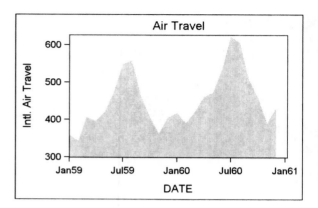

```
title "Air Travel";
proc sgplot data=sashelp.air;
   where date >= '01jan1959'd and date <='01jan1961'd;
   yaxis offsetmin=0 label="Intl. Air Travel";
   xaxis offsetmin=0 interval=semiyear tickvalueformat=monyy5.;
   band x=date upper=air lower=300;
   run;
```

Figure 8.4.4: The MINOR Option

By default, minor tick marks are not drawn on the axis.

The MINOR option turns them on for both TIME axes and LOG axes.

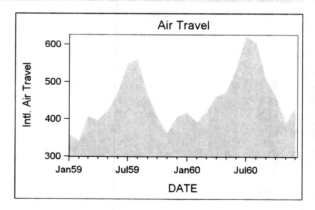

```
title "Air Travel";
proc sgplot data=sashelp.air;
   where date >= '01jan1959'd and date <='01jan1961'd;
   yaxis offsetmin=0 label="Intl. Air Travel";
   xaxis offsetmin=0 interval=semiyear tickvalueformat=monyy5. minor;
   band x=date upper=air lower=300;
   run;
```

8.5 Discrete Axis

Figure 8.5.1: Fit Policy

When collisions occur on a horizontal discrete axis, the FITPOLICY option can be used to control which policy or sequence of policies are used to resolve the collision.

In this example, the STAGGER policy was used instead of the ROTATE default.

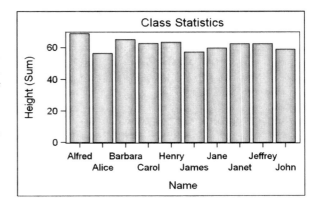

```
title "Class Statistics";
proc sgplot data=sashelp.class (obs=10);
  xaxis fitpolicy=stagger;
  vbar name / response=height dataskin=pressed;
  run;
```

Figure 8.5.2: Discrete Order

By default, the tick values of categorical charts and box plots are sorted by ascending unformatted value. The DISCRETEORDER option can be used to change the sort order.

In this example, the bars are drawn in DATA order. For bar charts, line charts, and dot plots, the CATEGORYORDER option can be used to sort by response values.

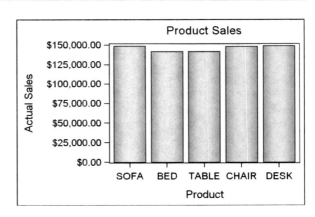

```
title "Product Sales";
proc sgplot data=sashelp.prdsale;
  xaxis discreteorder=data;
  vbar product / response=actual dataskin=pressed;
  run;
```

8.6 Legends

Figure 8.6.1: Automatic Legends

When grouped plots or multiple plots are specified, the SG procedures evaluate the request to determine the best legend to display.

Only one automatic legend is drawn. If more legends are needed, you need to create custom legends (see Figure 8.6.2).

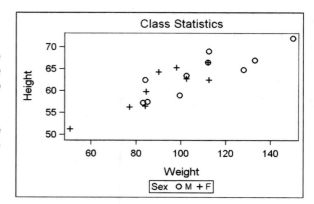

```
Title "Class Statistics";
proc sgplot data=sashelp.class;
  scatter x=weight y=height / group=sex;
  run;
```

Figure 8.6.2: Custom Legends

To customize the default legend or to create your own, use the KEYLEGEND statement.

Assign names to the plots to be included in the legend(s) and list their names on the KEYLEGEND statement(s) in the order in which you want them displayed. The LEGENDLABEL plot option can be used to customize the legend label for a plot without groups.

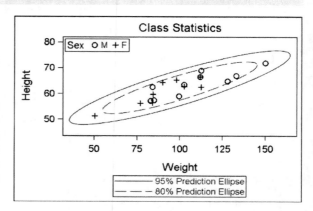

```
proc sgplot data=sashelp.class;
  scatter x=weight y=height / group=sex name="s";
  ellipse x=weight y=height /  name="e1";
  ellipse x=weight y=height / alpha=0.2  name="e2";
  keylegend "e1" "e2";
  keylegend "s" / location=inside title="Sex";
  run;
```

Figure 8.6.3: The LOCATION and POSITION Options

The LOCATION option specifies whether a legend is inside or outside of the data area. The POSITION option species one of eight positions in the legend's location.

If the legend is inside the data area, and POSITION is not specified, the legend will automatically position itself in an area with the least amount of data collision.

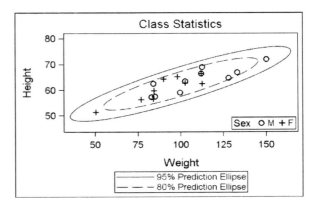

```
Title "Class Statistics";
proc sgplot data=sashelp.class;
   scatter x=weight y=height / group=sex name="s";
   ellipse x=weight y=height /  name="e1";
   ellipse x=weight y=height / alpha=0.2  name="e2";
   keylegend "e1" "e2";
   keylegend "s" / location=inside position=bottomright title="Sex";
   run;
```

Figure 8.6.4: The ACROSS and DOWN Options

The ACROSS and DOWN options control the layout of the legend entries. The ACROSS option limits how many legend entries are drawn across.

In this example, setting ACROSS=1 forces the legend entries to be drawn in one column.

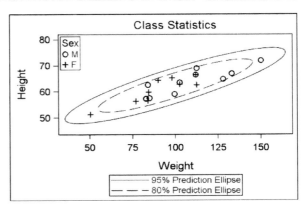

```
Title "Class Statistics";
proc sgplot data=sashelp.class;
   scatter x=weight y=height / group=sex name="s";
   ellipse x=weight y=height /  name="e1";
   ellipse x=weight y=height / alpha=0.2  name="e2";
   keylegend "e1" "e2";
   keylegend "s" / location=inside across=1 title="Sex";
   run;
```

8.7 Insets

Figure 8.7.1: Inset Types

This example shows two kinds of insets: a text inset and a table inset. Each quoted string in a text inset creates a new line. Each label/value pair in a table inset creates a new entry in the table.

Insets can be positioned in any of eight positions. However, if no position is specified, the inset automatically positions itself in an area with the least amount of data collision.

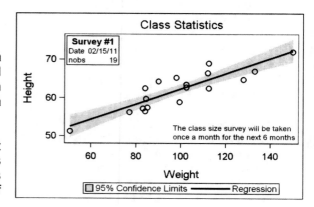

```
Title "Class Statistics";
proc sgplot data=sashelp.class;
   reg x=weight y=height / clm;
   inset ("Date"="02/15/11" "nobs"="19") / title="Survey #1" border
         textattrs=(size=7pt) titleattrs=(size=8 weight=bold);
   inset "The class size survey will be taken"
         "once a month for the next 6 months" / textattrs=(size=7pt);
run;
```

Figure 8.7.2: Rich-Text Inset

Not only do insets support Unicode characters like other strings in the SG procedures, but they also support superscripts and subscripts.

Use the ODS ESCAPECHAR option to insert rich text into your inset strings.

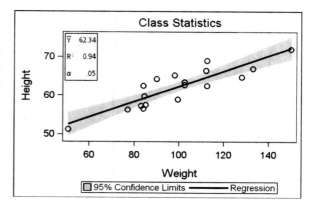

```
ods escapechar='~';
proc sgplot data=sashelp.class;
   reg x=weight y=height / clm;
   inset ( "Y~{unicode bar}"="62.34" "R~{sup '2'}"="0.94"
         "~{unicode alpha}"=".05" ) / border textattrs=(size=7pt);
run;
```

Chapter 9

Annotation and Attribute Maps (SAS 9.3)

9.1 Annotation 235
9.2 Attribute Maps (9.3) 251

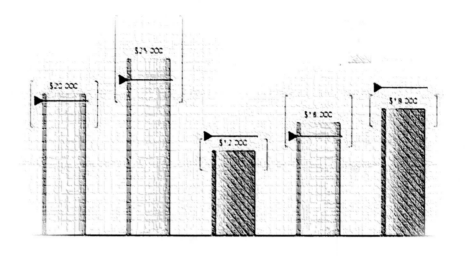

"Only two things are infinite, the universe and human stupidity, and I'm not sure about the former."
~ Albert Einstein

Chapter 9: Annotation and Attribute Maps (SAS 9.3)

In SAS 9.3, two new features were added to the SG procedures that are driven by auxiliary data sets. The first feature is an annotation facility that enables you to draw text and other graphical primitives on your graph output to supplement or enhance the plot information. The second feature is called an *attribute map*, which enables you to define the visual attributes of particular group values.

9.1 Annotation

The basic idea for annotation is that you can define a data set that contains the custom drawing actions you want to perform on a graph (see Figure 9.1.1). The data set has reserved column names, each providing the specific information needed to create the graphical primitives. The name of this data set is provided on the procedure statement using the SGANNO option.

Each observation draws one graphical primitive on the graph as defined by the FUNCTION column. If the function is a POLYGON or a POLYLINE, then this observation, together with subsequent POLYCONT values, defines the vertices of the figure. The information needed for each function is provided in the other named columns of the data set. Figure 9.1.2 shows the program code needed to create the graph shown in Figure 9.1.3.

Function	drawspace	x1	x2	yc1	yc2
Polyline	datavalue	2.3		Chrysler	
Polycont	datavalue	3.5		Chrysler	
Polycont	datavalue	3.5		Fiat	
Polycont	datavalue	2.8		Fiat	
Polyline	datavalue	3.5		Suzuki	
Polycont	datavalue	4.5		Suzuki	
Arrow	datavalue	4.5	4.5	Suzuki	Hyundai

Figure 9.1.1

```
title "Top Global Automakers...";
proc sgplot data=autos sganno=anno;
    hbarparm category=Automaker
      response=million_units / <options>;
    yaxis display=(nolabel) reverse;
run;
```

Figure 9.1.2

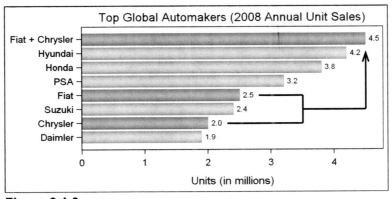

Figure 9.1.3

When you create the annotation data set using DATA step code, be sure to check the length of your columns. This annotation definition uses a number of keywords of varying lengths. You might start with an attribute that is only four characters wide and change it to something eight characters wide later in the DATA step. The eight-character value will be truncated to four characters unless you use the LENGTH statement or set a column range in the INPUT statement. If you see odd results in your graph, be sure to check the log to see if invalid attributes are specified due to truncation.

Many SAS/GRAPH users will be familiar with the annotation concepts described in this chapter, as they are based on the SAS/GRAPH annotation facility. However, this new facility has been redesigned to take advantage of the capabilities of the ODS Graphics system, such as transparency and rich text support. The data set column names and values have been defined in a way to help make them more memorable and self-documenting. The facility uses various drawing spaces in non-paneled plots to help simplify the placement of graphics primitives.

For the SAS 9.3 release, the following functions are supported in the FUNCTION column:
- TEXT – Used to draw a text string in the graph.
- TEXTCONT – Used to continue a text string from a previous TEXT or TEXTCONT function. This function is typically to perform rich text operations, such as changing color or font attributes.
- IMAGE – Used to place an image into the graph.
- LINE – Used to draw a line segment in the graph.
- ARROW – Used to draw an arrow in the graph.
- RECTANGLE – Used to draw rectangles or squares in the graph.
- OVAL – Used to draw ovals or circles in the graph.
- POLYGON – Used to draw a closed line figure in the graph. Because this is a closed figure, the figure may be filled. This operation specifies the starting point of the polygon. Each subsequent POLYCONT operation is used to draw one line segment from the previous ending point.
- POLYLINE – Used to draw an open line figure in the graph. This operation specifies the starting point of the polygon. Each subsequent POLYCONT operation is used to draw one line segment from the previous ending point.
- POLYCONT – Used to specify the next drawing point for a POLYGON or POLYLINE figure. All visual attributes for the figure must be specified on the initial POLYGON or POLYLINE statement.

An important concept to understand when working with annotations is the concept of drawspace. The drawspace specification is a combination of both an area and a unit. Not all areas are supported by all SG procedures. The SGPLOT procedure supports all areas, while the SGPANEL and SGSCATTER procedures support only the graph and layout areas. The graph below shows where these areas are located in a typical SGPLOT output (Figure 9.1.4).

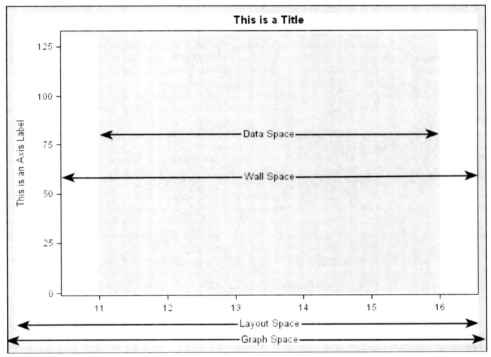

Figure 9.1.4

The units are either PERCENT or PIXEL, except the DATA area which also supports a VALUE unit. As an example, if you wanted to draw some annotation in the wall area using percentages, you could use the WALLPERCENT drawspace. Currently, annotations must be drawn using absolute coordinates; there is not a relative drawing mode.

Annotations do not reserve space in the graph. If you need to reserve space in the graph for an annotation, there are two techniques you can use. If you need space outside of the plot area, the SG procedures support a new option called PAD that gives you the ability to padding to any of the four edges of the graph. To reserve space within the plot area, use the OFFSETMIN and OFFSETMAX options on the axis statements.

Figure 9.1.5: Creating Axis Tick Values with Unicode

This example uses a user-defined format to create age ranges for each box. To show these ranges in the most compact way, you can use a Unicode "≤" in the tick value. To add these characters, you can turn off the tick values and use annotation to create the tick values. The ODS ESCAPECHAR functionality can be used to add Unicode, superscripts, or subscript to your annotation strings. The data set contains a value for each range so that the values can be positioned in data space. The label uses the Unicode function to generate the "≤" ('2264'x). The new PAD option is used to reserve space for the annotated axis values.

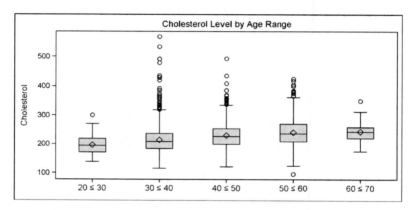

function	y1space	x1space	y1	width	x1	label
text	graphpercent	datavalue	7	15	21	20 ≤ 30
text	graphpercent	datavalue	7	15	31	30 ≤ 40

```
data anno;
retain function 'text' y1space 'graphpercent' x1space 'datavalue' y1 7 width 15;
input x1 label $ 4-33;
cards;
21 20 (*ESC*){unicode '2264'x} 30
31 30 (*ESC*){unicode '2264'x} 40
41 40 (*ESC*){unicode '2264'x} 50
51 50 (*ESC*){unicode '2264'x} 60
61 60 (*ESC*){unicode '2264'x} 70
;
run;

ods graphics / reset=index width=6in height=3in;
Title1 "Cholesterol Level by Age Range";
proc sgplot data=sashelp.heart sganno=anno pad=(bottom=8%);
  format AgeAtStart agefmt.;
  xaxis display=(nolabel novalues);
  vbox cholesterol / category=AgeAtStart;
run;
```

Figure 9.1.6: Creating Multi-line Axis Tick Values

When tick values contain multiple words, it is sometimes more space-efficient to split the value into multiple lines. There is a technique using annotation that can help you create these multi-line values. The key is using the WIDTH column to specify a text width. When a text string reaches the specified width, the system will automatically try to wrap the string. By combining the correct width with the "center" value on the JUSTIFY column, you can create center-justified, multi-line tick values.

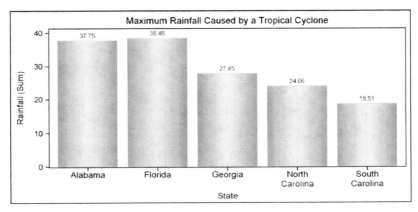

xc1	x1space	y1space	anchor	function	width	justify	y1	label
North Carolina	datavalue	wallpercent	top	text	15	center	-0.5	North Carolina
South Carolina	datavalue	wallpercent	top	text	15	center	-0.5	South Carolina

```
data anno;
set sgbook.rainfall (keep=state) end=_last_;
length x1space $ 11 y1space $ 13 anchor $ 6;
rename state=xc1;
retain function "text" x1space "datavalue" y1space "wallpercent" width 15
justify "center" y1 -0.5 anchor "top";
label=state;
output;
if (_last_) then do;
   x1space = "wallpercent";
   y1space = "graphpercent";
   anchor="bottom";
   x1=50;
   y1=1;
   label="State";
   output;
end;

title "Maximum Rainfall Caused by a Tropical Cyclone";
proc sgplot data=rainfall sganno=anno
            pad=(bottom=15%) noautolegend;
   xaxis display=(nolabel novalues);
   vbar state / response=rainfall dataskin=pressed
        datalabel;
run;
```

Figure 9.1.7: Creating an Axis-aligned Statistics Table

This graph contains an axis-aligned statistics table of class weight and height. The Y-coordinates and the table values are extracted from the bar chart's input data set. After the last value is processed (_last_ is set), we add the column headers for the table. The PROC SGPLOT call to make the bar chart uses the new PAD option to provide space for the table.

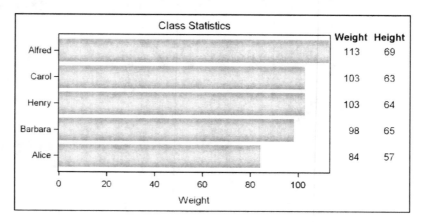

Name	Height	Weight	x1space	y1space	anchor	label	function	yc1	x1
Alfred	69.0	112.5	wallpercent	datavalue	Right	113	text	Alfred	111.0
Alfred	69.0	112.5	wallpercent	datavalue	Right	69	text	Alfred	122.0

```
data anno;
set sashelp.class (obs=5 keep=name weight height) end=_last_;
length x1space $ 11 y1space $ 11 anchor $ 8 label $ 6;
retain function 'text' x1space 'wallpercent' y1space 'datavalue' anchor 'right';
yc1=name; /* Y-coordinate is the name */
x1=111; /* percent beyond the edge of the wall */
label=put(weight, F3.0);
output;
x1=122; /* percent beyond the edge of the wall */
label=put(height, F3.0);
output;
if (_last_) then do; /* Add table headers to the end of the data */
   y1space = 'wallpercent';
   width = 20;
   anchor = 'top';
   textweight='bold';
   y1 = 103;
   x1=107.5;
   label="Weight";
   output;
   x1=119.5;
   label="Height";
   output;
```

```
title "Class Statistics";
proc sgplot data=sashelp.class (obs=5) sganno=anno
   pad=(right=40%);
   yaxis display=(nolabel);
   hbar name / response=weight categoryorder=respdesc
               dataskin=pressed nostatlabel;
run;
```

Figure 9.1.8: Creating a Simple Forest Plot with Annotation

This example combines the techniques of data alignment and text width in the previous examples to create a simple forest plot. Notice how the effect labels along the bottom axis are positioned based on wall position, and the width keeps long labels from colliding. If you know you are going to have long effect labels, you may want to specify more bottom padding in the SGPLOT procedure. Also see Figure 12.2 for an alternate way to create a forest plot without annotation.

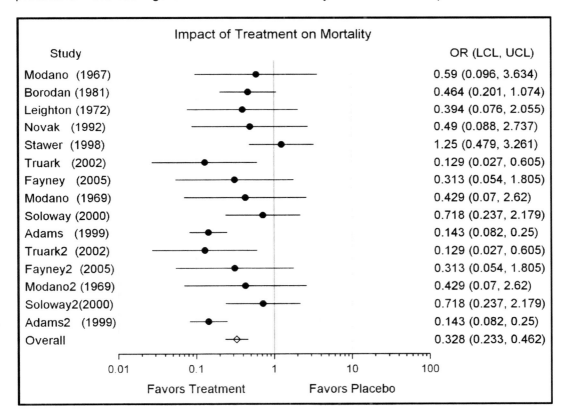

Study	Summary	label	anchor	y1space	x1space	function	width	textsize	yc1	x1
Modano (1967)	0.590 (0.096, 3.634)	Modano (1967)	left	datavalue	graphpercent	text	50	8	Modano (1967)	1
Modano (1967)	0.590 (0.096, 3.634)	0.590 (0.096, 3.634)	left	datavalue	wallpercent	text	50	8	Modano (1967)	101

```
/* Annotation for the study column, the stat column, */
/* and the effect labels.                            */
data anno;
set sgbook.labels end=_last_;
length label $ 50 anchor $ 6 y1space $ 13 x1space $ 12;
retain function "text" y1space "datavalue" width 50 anchor "left" ;
yc1=study;
label=study;
x1space="graphpercent";
x1=1;
output;
label=summary;
x1space="wallpercent";
x1=101;
output;
if (_last_) then do;
  /* Add the effect labels */
  yc1=" ";
  x1space="wallpercent";
  y1space="graphpercent";
  anchor="bottom";
  y1=2;
  x1=25;
  label="Favors Treatment";
  output;
  x1=75;
  label="Favors Placebo";
  output;
  /* Add the column headers */
  y1space="wallpercent";
  x1space="graphpercent";
  y1=100;
  x1=9;
  label="Study";
  output;
  x1=89;
  label="OR (LCL, UCL)";
  output;
end; run;
proc sgplot data=sgbook.meta nocycleattrs noautolegend sganno=anno
    dattrmap=attrmap pad=(left=22% right=28% bottom=7%);
  yaxis reverse display=none;
  xaxis type=log logbase=10 min=.01 max=100 minor display=(nolabel)
      offsetmin=0 offsetmax=0;
  refline 1 / axis=x;
  refline .01 .1 10 100 / axis=x lineattrs=GraphGridLines;
  highlow y=study high=UpperCL low=LowerCL;
  scatter y=study x=oddsratio / attrid=a group=study_type;  run;
```

Figure 9.1.9: Adding a Company Logo

You can create a small annotate data set to add a company logo to all of the graphs in your report, assuming you want the logo in the same location. A typical place to put this logo is in one of the extreme corners of the graph. The key to using the data set is to specify the proper padding in the SG procedure to prevent collisions with the plot axes or other graph features.

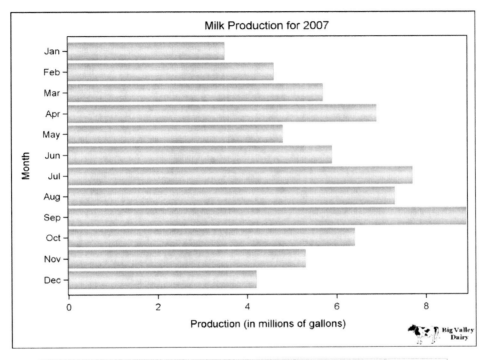

function	anchor	x1	y1	width	drawspace	image
Image	bottomright	100	0.5	20	graphpercent	logo.png

```
data anno;
retain function "Image" anchor "bottomright" x1 100 y1 0.5
       width 20 drawspace "graphpercent" image "logo.png";
run;

title "Milk Production for 2007";
proc sgplot data=sgbook.milk_production pad=(bottom=5%) sganno=anno;
  yaxis discreteorder=data;
  hbar month / response=production dataskin=pressed;
run;
```

Figure 9.1.10: Background Images in the Graph

In this example, the IMAGESCALE column is used to tile the texture image. The WIDTH and HEIGHT columns define the size of the area in which to tile the image. The ANCHOR, X1, and Y1 columns determine where the tiling area is positioned. The LAYER column is used to push the image to the back of the graph, even behind the wall of the plot area.

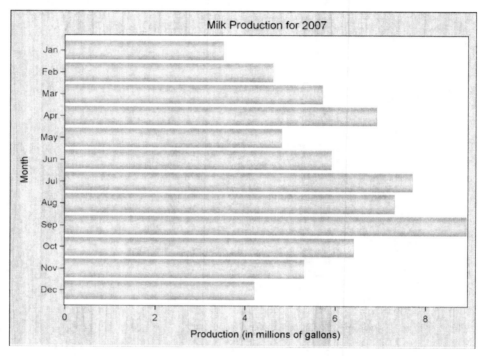

function	anchor	x1	y1	width	height	widthunit	heightunit	imagescale	drawspace	transparency	layer	Image
Image	bottomright	100	0.5	640	480	pixel	pixel	tile	graphpercent	0.5	back	graywood.jp

```
data anno;
retain function "Image" anchor "bottomright" x1 100 y1 0.5 width 640
    height 480 widthunit 'pixel' heightunit 'pixel' imagescale 'tile'
    drawspace "graphpercent" layer 'back' transparency 0.5
    Image "graywood.png";
run;
```

Figure 9.1.11: Images as Curve Labels

In this example, an image is used to label each curve instead of a text label. All images are positioned just outside of the x-axis data area. The Y position of the images comes from the Y value of the plot points at the year 2000. The ability to position images in data space also opens up the possibility of using images as plot points. Note that annotations do not reserve space; therefore, you must provide an axis offset large enough to contain the images within the wall area.

In addition to the image labels, the company logo and the footnote text are also annotated. The text is annotated because the text from a FOOTNOTE statement would have been raised up with plot when adding padding for the company logo.

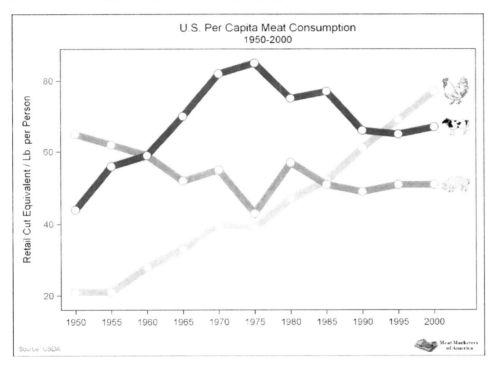

x1space	y1space	anchor	width	widthunit	function	x1	y1	image	textsize	label
datapercent	datavalue	Left	40	pixel	image	102	77	chicken.jpg	.	
datapercent	datavalue	left	40	pixel	image	102	67	cow.jpg	.	
datapercent	datavalue	left	40	pixel	image	102	51	pig.jpg	.	
graphpercent	graphpercent	bottomright	90	pixel	image	99	1	Logo.png	.	
graphpercent	graphpercent	bottomleft	150	pixel	text	1	1	Logo.png	6	Source: USDA

```
data anno;
length x1space $ 13 y1space $ 13 anchor $ 11;
/* Query for the observation at year 2000 */
set meat_consumption (where=(year='01jan2000'd)) end=_last_;
retain anchor 'left' y1space 'datavalue' x1space 'datapercent'
       width 40 widthunit 'pixel' function 'image' x1 102;
y1 = chicken;
image = chicken.jpg";
output; /* Chicken image */
y1 = beef;
image = cow.jpg";
output; /* Cow image */
y1 = pork;
image = pig.jpg";
output; /* Pig image */
if (_last_) then do;
   /* The company logo */
   x1space = "graphpercent";
   y1space = "graphpercent";
   anchor = "bottomright";
   x1 = 99;
   y1 = 1;
   width=90;
   image = Logo.png";
   output;
   /* The footnote text */
   function = "text";
   anchor = "bottomleft";
   x1 = 1;
   width=150;
   textsize = 6;
   label = "Source: USDA";
   output;
end;   run;
```

Figure 9.1.12: Using Arrows to Emphasize Data

In this example, arrows are used to emphasize outlying values in the plot. Notice how a common X2 and Y2 value are use to draw the arrows to the same point. The single label is added after all of the data is processed.

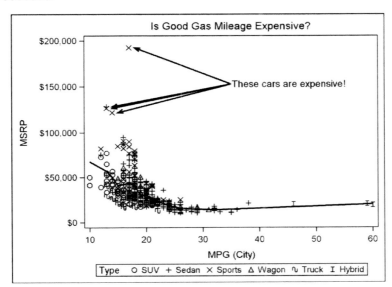

y1	x1	x1space	y1space	function	x2space	y2space	x2	y2	direction	scale
$128,420	14	datavalue	datavalue	Arrow	wallpercent	wallpercent	50	75	in	0.5
$121,770	15	datavalue	datavalue	Arrow	wallpercent	wallpercent	50	75	in	0.5

```
data anno;
set sgbook.expensive (keep=msrp mpg_city) end=_last_;
length x1space $ 11 y1space $ 11;
retain function "Arrow" x1space "datavalue" y1space "datavalue" x2space "wallpercent"
       y2space "wallpercent" x2 50 y2 75 direction "in" scale 0.5;
rename msrp=y1 mpg_city=x1;
mpg_city = mpg_city + 1;
output;
if (_last_) then do;
  function = "text";
  x1space = "wallpercent";
  y1space = "wallpercent";
  mpg_city=50;
  msrp=75;
  anchor="left";
  width=50;
  label="These cars are expensive!";
  output; end;
```

```
Title "Is Good Gas Mileage Expensive?";
proc sgplot data=sashelp.cars sganno=anno;
  scatter x=mpg_city y=msrp / group=type;
  loess x=mpg_city y=msrp / nomarkers;
run;
```

Figure 9.1.13: Making a Bubble Legend

This legend is constructed using four annotation functions: RECTANGLE, OVAL, LINE, and TEXT. The minimum and maximum size of the annotated bubbles is synchronized with the bubble plot by using the BRADIUSMIN and BRADIUSMAX options.

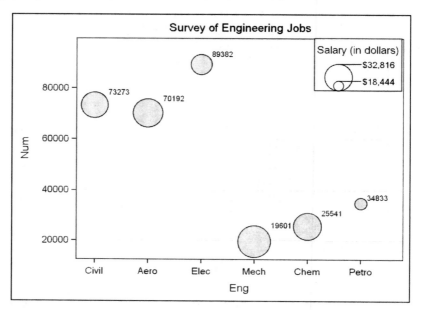

drawspace	widthunit	heightunit	linethickness	textsize	function	x1	y1	width	height	anchor
wallpercent	pixel	pixel	1	12	Rectangle	86	76.5	140	87	bottom
wallpercent	pixel	pixel	1	12	Oval	80	76.5	44	44	bottom

```
data anno;
retain drawspace "wallpercent" widthunit "pixel" heightunit "pixel"
      linethickness 1 textsize 8;
length function $ 9;
input function $ x1 y1 width height x2 y2 textsize anchor $ label $ 48-66;
cards;
Rectangle  86 76.5  140   87    .    .   12 bottom
Oval       80 76.5   44   44    .    .   12 bottom
Oval       80 76.5   16   16    .    .   12 bottom
Line       80 87.9    .    .   87  87.9  12 bottom
Line       80 80.5    .    .   87  80.5  12 bottom
Text       86 98    140    .    .    .   12 top      Salary (in dollars)
Text       87 88.1  140    .    .    .    8 left     $32,816
Text       87 80.4  140    .    .    .    8 left     $18,444;      run;
```

Figure 9.1.14: Making a Polyline Figure

In this example, a POLYLINE function is used with other functions to create an annotation showing the result of two companies combining into one. A group variable is used to color the bars differently from the rest of the companies.

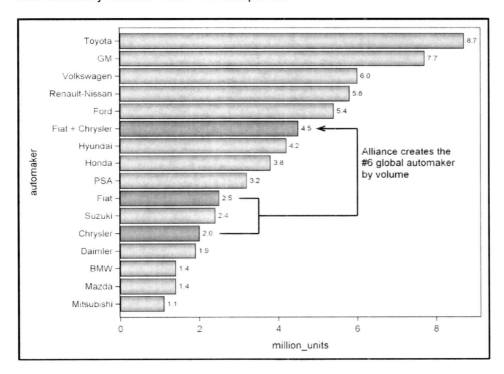

yc1	drawspace	function	x1	x2	yc2	anchor	width	label
Chrysler	datavalue	polyline	2.5	.			.	
Chrysler	datavalue	polycont	3.5	.			.	

```
data anno;
length yc1 $ 15;
retain drawspace "datavalue";
function="polyline";
yc1="Chrysler";
x1=2.5;
output;
function="polycont";
x1=3.5;
output;
yc1="Fiat";
output;
x1=3.0;
output;
function="polyline";
yc1="Suzuki";
x1=3.5;
output;
function="polycont";
x1=6.0;
output;
yc1="Fiat + Chrysler";
output;
function="arrow";
x2=5.0;
yc2="Fiat + Chrysler";
output;
function="text";
yc1="Honda";
x1=6.1;
anchor="Left";
width=30;
label="Alliance creates the #6 global automaker by volume";
output;
run;

proc sgplot data=sgbook.autos noautolegend sganno=anno;
  hbarparm category=automaker response=million_units / datalabel
        group=colorvar dataskin=pressed;
run;
```

9.2 Attribute Maps (9.3)

Attribute maps are used to assign visual attributes to specific group data values. Like annotations, this feature uses a data set to define the content of the feature; however, the definition is much smaller.

Required Column(s):

ID	Char	The ID of the attrmap. This value is referenced from the ATTRID option on the plot statements.
VALUE	Char	The data value to be assigned to the attributes. The keyword _OTHER_ in the column can be used to define the attributes of any values that are not explicitly defined in the map. Numeric values should represent their final formatted form.

Optional Column(s):

MARKERSTYLE	Char	A style element reference for marker attributes (e.g. GraphData1, GraphData2, etc.).
MARKERSYMBOL	Char	Either a literal marker symbol (circle, plus, etc.) or an attribute style reference (GraphData1:markersymbol).
MARKERCOLOR	Char	Either a literal marker color (red, cxff0000, etc.) or an attribute style reference (GraphData1:contrastcolor).
LINESTYLE	Char	A style element reference for line attributes (e.g. GraphData1, GraphData2, etc.).
LINEPATTERN	Char	Either a literal line pattern (solid, dash, etc.) or an attribute style reference (GraphData1:linepattern).
LINECOLOR	Char	Either a literal line color (red, cxff0000, etc.) or an attribute style reference (GraphData1:contrastcolor).
FILLSTYLE	Char	A style element reference for fill attributes (e.g. GraphData1, GraphData2, etc.).
FILLCOLOR	Char	Either a literal fill color (red, cxff0000, etc.) or an attribute style reference (GraphData1:color).

You can define multiple attribute maps within the same data set; however, make sure that the data set stays sorted by the ID column. The ID is referenced from the procedure's PLOT statement and the attributes applied to the GROUP variable. If there is not a GROUP variable on the PLOT statement, the attribute map is ignored. Also note that values in the attribute map are case-sensitive.

The optional columns are applied to the plot only if the attributes can be applied to the plot. Otherwise, the extra columns are ignored. If a group value is not found in the map, the attributes for that group will come from the ODS style.

Figure 9.2.1: Controlling the Appearance of Plot Lines

In this example, the attribute map is defined to control the appearance of both the plot lines and the markers for each group value.

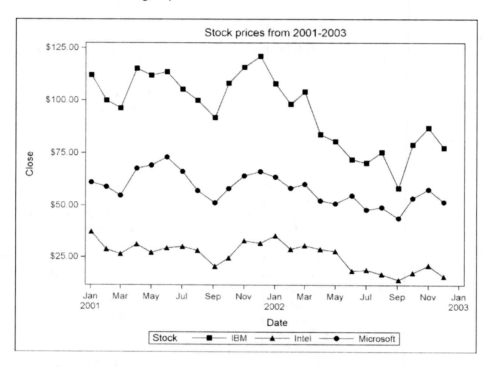

```
data attrmap;
length value $ 9 markersymbol $ 14;
retain id 'my_id' linepattern 'solid';
input value $ markersymbol;
cards;
Microsoft  circlefille
IBM        squarefilled
Intel      trianglefilled
;
run;
```

The DATA step produces an attribute map with the following information:

value	markersymbol	id	linepattern
Microsoft	circlefille	my_id	solid
IBM	squarefilled	my_id	solid
Intel	trianglefilled	my_id	solid

The value "my_id" in the ID column is referenced by the ATTRID option in the following code to identify which map in the data set to use. The values from the GROUP variable are compared to the map values in the VALUE column to determine if the defined attributes in the attribute columns should be applied.

```
ods listing style=journal;
Title1 "Stock prices from 2001-2003";
proc sgplot data=sashelp.stocks dattrmap=attrmap;
where Date >= '01jan2001'd and Date <= '01jan2003'd;
series x=Date y=close / markers group=stock attrid=my_id;
run;
```

Chapter 10

Classification Panels

10.1　Introduction　257
10.2　SGPANEL Procedure　258
10.3　PANELBY Statement　259
10.4　Classification Panels　261
10.5　Paging of Large Panels　266

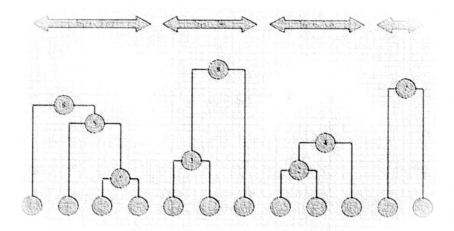

*"I have not failed. I've just found
10,000 ways that won't work."
~ Thomas Alva Edison*

Chapter 10: Classification Panels

In this chapter, we step into the world of multi-cell classification panels using the SGPANEL procedure. These graphs are valuable tools for understanding the data by displaying the data classified by one or more class variables.

Classification panel graphs allow us to examine the relationship between variables for all crossings of the class variables. Therefore, we can examine effects of a drug over time classified by the gender of the person in the study.

10.1 Introduction

In addition to the plots supported by the SGPLOT procedure, the SGPANEL procedure supports a PANELBY statement. This is the key difference between the two procedures. As mentioned in Section 2.3, the PANELBY statement allows us to specify the LAYOUT type, and the classification variables. The graph space is subdivided into smaller "cells" based on the number of levels for each class variable(s).

Each cell in the graph is populated with a subset of the data that corresponds to the class variable value(s). The plot statements in the PROC step after the PANELBY statement define the "prototype" or "rubber stamp" that is used to populate each cell, as shown in Figures 10.1.1 and 10.1.2 below.

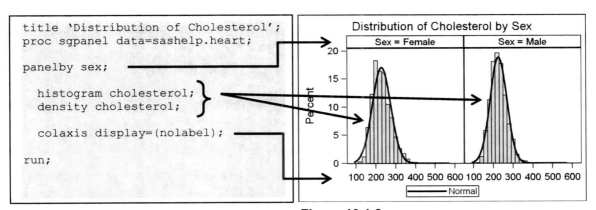

Figure 10.1.1   **Figure 10.1.2**

The key features of the example shown in Figures 10.1.1 and 10.1.2 are:
- The PANELBY variable is "Sex". Since the data has two values for sex, this results in two cells in the graph.
- Each cell is populated with a HISTOGRAM and a DENSITY statement for the associated data.
- Options on other statements such as COLAXIS apply to the whole graph.

Note: Each cell essentially contains a graph of the type created by the SGPLOT procedure. So, all the techniques that we have learned in the previous chapters apply here, with just a few exceptions.

10.2 SGPANEL Procedure

The SGPANEL procedure supports the following syntax:

```
Proc SGPANEL data=datasetname <options>;
```

Options:

CYCLEATTRS	boolean	Cycle through style elements for each plot
DATA	=sas-data-set	Optional data set
DESCRIPTION	=string	Description string
DATTRMAP	=sas-data-set	Data set defining an attribute map
NOAUTOLEGEND	boolean	Do not cycle style elements for each plot
NOCYCLEATTRS	boolean	Do not cycle style elements for each plot
PAD	=value	Padding around the outside of the graph
SGANNO	=sas-data-set	Data set containing the annotations
TMPLOUT	=string	File name for generated graph template code

The procedure supports the following statements:
- PANELBY
- Basic Plots: BAND, NEEDLE, SCATTER, STEP, SERIES, VECTOR, BUBBLE, HIGHLOW, HBARPARM, VBARPARM
- Fit Plots: LOESS, PBSPLINE, REG, LINEPARM
- Distribution Plots: DENSITY, HISTOGRAM, HBOX, VBOX
- Categorical Plots: DOT, HBAR, HLINE, VBAR, VLINE
- Other: KEYLEGEND, REFLINE, COLAXIS, ROWAXIS

The procedure has the following syntax:

```
PROC SGPANEL < DATA= data-set > < options >;
  PANELBY classvar1 < classvar2 ... <classvarN > < / options >;
  < plot-statement(s) >;
  < refline-statement(s) >;
  < axis-statement(s) >;
  < keylegend-statement(s) >;
RUN;
```

As seen in the examples in Chapter 3, one or more plot statements can be provided that will work together to create the graph. See the table of permissible combinations shown in section 2.4.

The syntax and features of some of these plots are very similar to the same statements in the SGPLOT procedure, with a few deviations related to positioning of data and curve labels. The syntax for these plot statements has been discussed in the preceding chapters as listed below, so we need not discuss them again here.
- Basic Plots   : Chapter 4
- Fit and Confidence Plots   : Chapter 5
- Distribution Plots   : Chapter 6
- Categorization Plots   : Chapter 7

The following statements are new to the SGPANEL procedure, or they behave differently than in the SGPLOT procedure. We will discuss these statements in detail in this chapter:
- PANELBY   : Defines the paneling structure of the graph
- KEYLEGEND   : Customizes the external legend
- REFLINE   : Adds common reference lines
- COLAXIS   : Customizes the common external column axis
- ROWAXIS   : Customizes the common external row axis

Uniform Data Ranges: For a panel graph, the axis ranges are always uniform as follows:
- All vertical axes in any one row always have a uniform data range.
- All horizontal axes in any one column always have a uniform data range.
- Vertical (or horizontal) data ranges may or may not be uniform between rows (or columns).
- A Y reference line is drawn at the same Y value in all cells of a row.
- An X reference line is drawn at the same X value in all cells of a column.

10.3 PANELBY Statement

The PANELBY statement is required and must be provided before any other plot, legend, axis, or refline statements. The syntax is:

```
panelby classvar1 <classvar2...classvarN> </ options>;
```

The space provided for the graph is subdivided into an even grid of cells based on the number of crossings for the class variables.

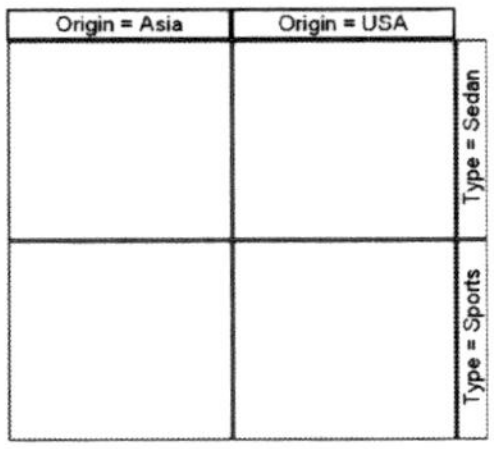

Figure 10.3.1

Four different types of layouts are available. These are panel (default), lattice, columnlattice and rowlattice as described below.

An example of the lattice layout is shown in figure 10.3.1, where the first class variable is used as the column variable and the second as the row variable.

Required Data Roles:

	column-list	classification variable(s)

Options:

COLHEADERPOS	=keyword	TOP | BOTTOM | BOTH
COLUMNS	=value	Number of columns
LAYOUT	=keyword	PANEL, LATTICE, ROWLATTICE, etc.
ROWHEADERPOS	=keyword	LEFT | RIGHT | BOTH
ROWS	=value	Number of rows
SPACING	=value	Spacing between cells
START	=keyword	TOPLEFT | BOTTOMLEFT
UNISCALE	=keyword	ALL | COLUMN | ROW

Boolean Options:

BORDER | NOBORDER	Displays borders
MISSING	Includes missing as a separate level
NOVARNAME	Displays only the class value in the header
ONEPANEL	Prevents paging of the panel

Layout Types – PROC SGPANEL supports four different layouts. Each layout subdivides the available graph region into a regular grid of cells. The layout of the cells and the location of the headers are different for each layout type as shown in Figure 10.3.2 below.

Each layout has its strengths as follows:
- Panel (Default) :Supports multiple class variables and is optimal for sparse data.
- Lattice   : Supports column and row variables and similar layout.
- ColumnLattice   : Supports one class variable to create a Lattice of columns (one row).
- RowLattice   : Supports one class variable to create a aLattice of rows (one column).

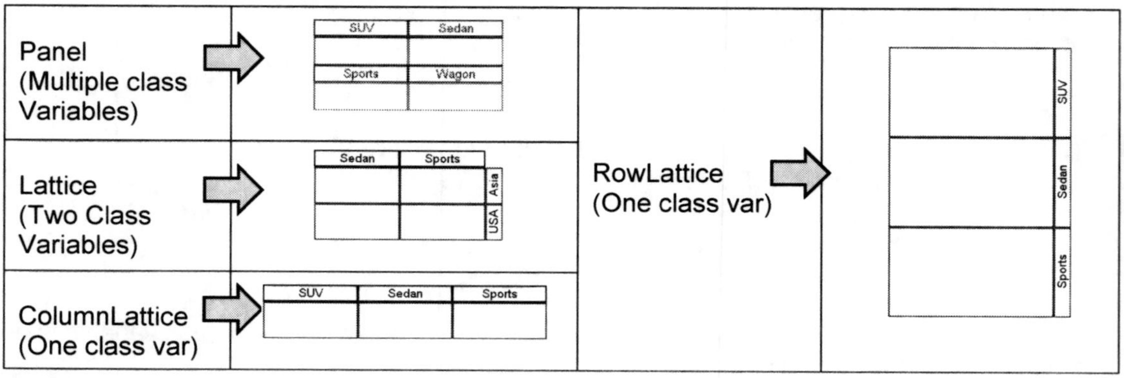

Figure 10.3.2

10.4 Classification Panels

Figure 10.4.1: Mileage by Horsepower and Type

This graph shows mileage by horsepower for vehicles classified by type. Having separate cells for each class allows us to plot the regression and confidence bands.

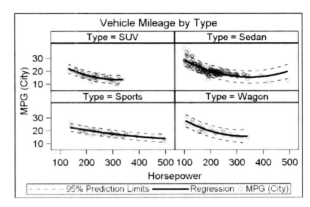

```
title 'Vehicle Mileage by Type';
proc sgpanel data=sashelp.cars;
   where (type='Sedan' or type='Sports' or type='SUV' or type='Wagon');
   panelby type;
   reg x=horsepower y=mpg_city / nomarkers cli degree=2;
   scatter x=horsepower y=mpg_city / transparency=0.7;
   run;
```

Figure 10.4.2: Mileage by Type

This graph shows the city and highway mileage for vehicles by type and origin.

The number of columns is set to 3, or else the procedure will attempt to create a balanced 2x2 grid.

The reference line at Y=20 is shown in all the cells. NOVARNAME suppresses the display of column labels in the cell headers.

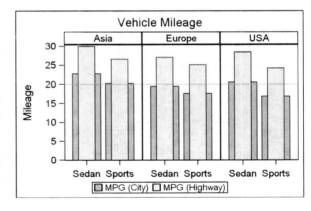

```
title 'Vehicle Mileage';
proc sgpanel data=sashelp.cars;
   where type='Sedan' or type='Sports';
   panelby origin / novarname columns=3;
   vbar type / response=mpg_city stat=mean fillattrs=graphdata7;
   vbar type / response=mpg_highway barwidth=0.5  stat=mean
       fillattrs=graphdata11;
   refline 20;
   colaxis grid display=(nolabel); rowaxis grid label='Mileage';
   run;
```

Figure 10.4.3: Panel by Origin and Type

This graph shows a panel with two class variables, Origin and Type. Each cell has a cell header for each of the class variables.

The space available for display of the data is quickly consumed by the cell headers.

This layout (panel) allows multiple class variables; however, the space is quickly consumed by the cell headers.

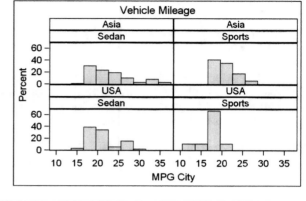

```
title 'Vehicle Mileage';
proc sgpanel data=sgbook.cars;
   where origin='USA' or origin='Asia';
   panelby origin type / novarname;
   histogram mpgc;
   colaxis grid; rowaxis grid;
   run;
```

Figure 10.4.4: Lattice by Origin and Type

This graph also displays a panel with two class variables, Origin and Type. Here we have used LAYOUT=Lattice, so Origin is used as the column variable and Type is used as the row variable.

Each row and column has a header, with the value of the class variable. This layout uses less space for headers.

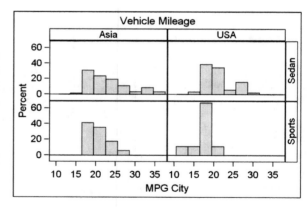

```
title 'Vehicle Mileage';
proc sgpanel data=sgbook.cars;
   where origin='USA' or origin='Asia';
   panelby origin type / layout=lattice novarname;
   histogram mpgc;
   colaxis grid; rowaxis grid;
   run;
```

Figure 10.4.5: Panel by Type

To create a one-column panel of horsepower by type, we can use the default layout (panel). To ensure one column, set COLUMN=1. We used option ONEPANEL to prevent paging of the graph into multiple images.

Each cell gets a cell header at the top, taking up a significant amount of the available space. See Figure 10.4.6 for an alternative layout.

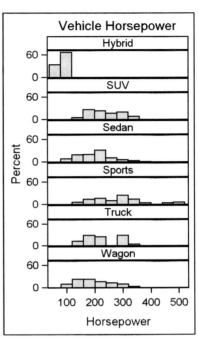

```
title 'Vehicle Horsepower';
proc sgpanel data=sashelp.cars;
   panelby type / onepanel novarname
         columns=1;
   histogram horsepower;
   colaxis grid; rowaxis grid;
   run;
```

Figure 10.4.6: Row Lattice by Type

In this graph, we used LAYOUT= RowLattice. This layout creates a lattice of rows (with one column). Each row gets a row header on one side (similar to the lattice layout).

Sometimes this layout is preferred when there are many short cells. The cell header goes on the side and results in an efficient usage of the graph space.

Note: All y-axis data ranges are uniform.

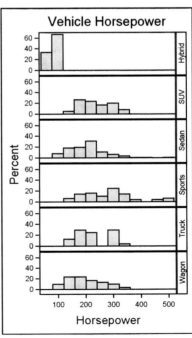

```
title 'Vehicle Horsepower';
proc sgpanel data=sashelp.cars;
   panelby type / layout=rowlattice onepanel
         novarname;
   histogram horsepower;
   colaxis grid; rowaxis grid;
   run;
```

Figure 10.4.7: Panel by Type

This is a graph of the vital signs for a patient in a clinical study. The vital signs are plotted by day of visit.

The range of values for each vital sign is different, from blood pressure, to weight (kg), to temperature (C). In such a use case, it is useful for each row to have its own data range. To get this, we set UNISCALE=Column.

```
title 'Vital Signs';
proc sgpanel data=sgbook.vitalsigns;
  where vstestcd <> 'HEIGHT';
  panelby vstestcd / layout=rowlattice onepanel
         novarname uniscale=column;
  series x=visitdy y=vsstresn;
  colaxis grid; rowaxis grid display=(nolabel);
run;
```

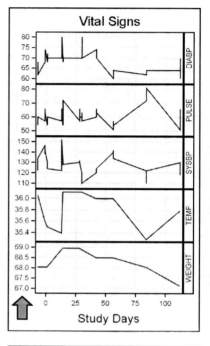

Figure 10.4.8: Row Lattice by Type

Here we have added the display of the normal ranges for some of the vital signs. These bands are different for each cell. This is a common feature needed to indicate if the measurements are within normal range.

To do this, we have added two new columns in the data with values for systolic and diastolic pressure. A band plot is added to display the normal range. If needed, data driven REFLINEs can also be added.

```
title 'Vital Signs';
proc sgpanel data=sgbook.vitalsigns;
  where vstestcd <> 'HEIGHT';
  panelby vstestcd / layout=rowlattice onepanel
         novarname uniscale=column;
  band x=visitdy lower=low upper=high / fill
       nooutline transparency=0.6;
  series x=visitdy y=vsstresn;
  colaxis grid; rowaxis grid display=(nolabel);
run;
```

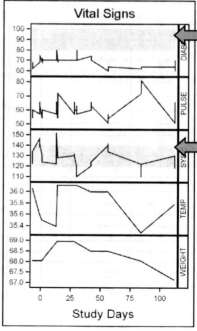

Figure 10.4.9: Column Lattice

This graph shows a column lattice, which is a lattice of multiple columns in one row.

```
title 'Vehicle Mileage by Origin ...';
proc sgpanel data=sashelp.cars
             noautolegend;
  where type <> 'Hybrid';
  panelby origin / layout=columnlattice
          onepanel novarname;
  vbar  type / response=mpg_city
        stat=mean dataskin=gloss;
  colaxis grid;
  rowaxis grid display=(nolabel);
run;
```

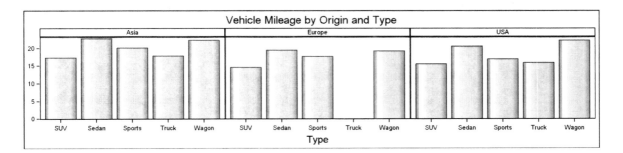

Figure 10.4.10: Grouped Bar Chart

SG procedures with SAS 9.2 do not support adjacent "cluster" grouped bar charts. This graph type can be simulated using the Layout=ColumnLattice option and other appropriate options as shown here.

SAS 9.3 does support "cluster" grouped bar charts as discussed earlier.

```
title 'Vehicle Mileage by Origin ...';
proc sgpanel data=sashelp.cars;
  where type <> 'Hybrid';
  panelby origin / onepanel novarname
     layout=columnlattice noborder
     colheaderpos=bottom;
  vbar  type / response=mpg_city
    stat=mean dataskin=gloss group=type;
  colaxis grid display=none
     offsetmin=0.2 offsetmax=0.2;
  rowaxis grid;
run;
```

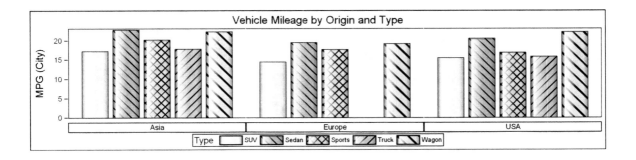

10.5 Paging of Large Panels

Figure 10.5.1: Layout Panel

This graph is a panel classified by origin and type. There are eight combinations, and we have specified a 2x2 grid.

The procedure creates two graph outputs to cover all the combinations. Note the Europe-Truck combination is completely dropped (including headers).

```
title 'Vehicle Mileage by Origin ...';
proc sgpanel data=sashelp.cars;
  where (origin='USA' or
      origin='Europe') and
      (type='Sedan' or type='Sports' or
      type='SUV' or type='Truck');
  panelby origin type / columns=2
      rows=2;
  histogram mpg_city;
  run;
```

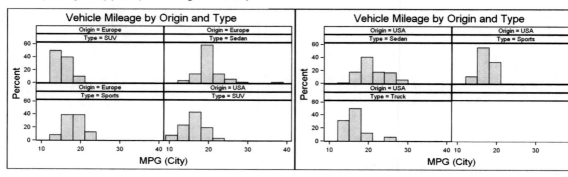

Figure 10.5.2: Layout Lattice

This graph is the same as Figure 10.5.1, except LAYOUT=Lattice.

In this case, all combinations are shown (including headers). However, cells with missing data are empty.

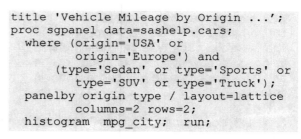

```
title 'Vehicle Mileage by Origin ...';
proc sgpanel data=sashelp.cars;
  where (origin='USA' or
      origin='Europe') and
      (type='Sedan' or type='Sports' or
      type='SUV' or type='Truck');
  panelby origin type / layout=lattice
      columns=2 rows=2;
  histogram mpg_city;  run;
```

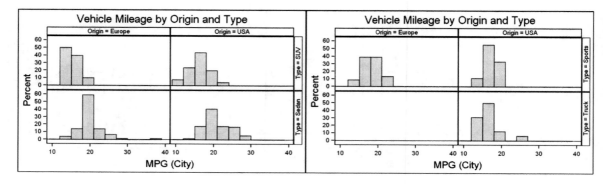

Chapter 11

Comparative and Matrix Plots

11.1   Introduction   269
11.2   SGSCATTER Procedure   269
11.3   PLOT Statement   270
11.4   COMPARE Statement   276
11.5   MATRIX Statement   281

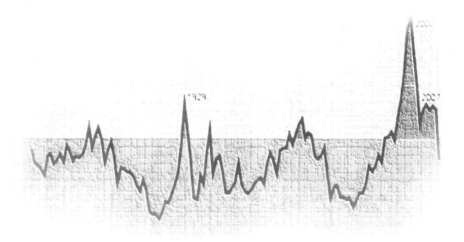

"If you don't know where you are going, you will wind up somewhere else."
~ Yogi Berra

Chapter 11: Comparative and Matrix Plots

In this chapter, we continue our journey through the world of multi-cell graphs. However, instead of creating graphs classified by one or more class variables, we will look into grids of plots that provide quick comparative views of the data.

11.1 Introduction

Comparative and matrix graphs primarily consist of scatter plots of multiple response variables or measures in the data set. These graphs are especially valuable for understanding the raw data received from the field. Often the analyst would like to get a feel for the data and find relationships between variables of a data set using "visual" means. This can be a valuable tool in the pre-analysis phase of the project.

The SGSCATTER procedure is specifically designed for the creation of grids of scatter plots in the form of comparative or matrix graphs. Now, you certainly can use the SGSCATTER procedure to create a single-cell scatter plot with fit curves and confidence bands. However, you will be better off using the SGPLOT procedure to create the single-cell graph as seen in the many examples shown in previous chapters. The SGSCATTER procedure is better suited for creation of grids of scatter plots with a few fit and confidence plots.

11.2 SGSCATTER Procedure

This procedure is very different from the SGPLOT and SGPANEL procedures. The syntax for this procedure is much more like the syntax for the GPLOT procedure and is as follows:

```
proc sgscatter <options>;
  plot or compare or matrix statement;
run;
```

Options:

DATA	=sas-data-set	Optional data set
DESCRIPTION	=string	Description string
DATTRMAP	=sas-data-set	Data set defining an attribute map
PAD	=value	Padding around the outside of the graph
SGANNO	=sas-data-set	Data set containing the annotations
TMPLOUT	=string	File name for generated graph template code

The SGSCATTER procedure supports the following three independent statements:

- PLOT – Creates multi-plot panels
- COMPARE – Creates comparative panels
- MATRIX – Creates square scatter plot matrices

11.3 PLOT Statement

The PLOT statement creates regular grids of scatter plots based on the "plot request". The syntax for this statement is as follows:

```
plot plot-request </ options>;
```

Plot request is in one of the forms listed below. A regular grid of scatter plots is created based on the plot request.
- y * x : creates a single scatter plot.
- y1 * x1 y2 * x2 : creates a graph with two scatter plots as defined.
- (y1 y2) * x : creates a graph with two separate scatter plots y1 * x and y2 * x.
- y * (x1 x2) : creates a graph with two separate scatter plots y * x1 and y * x2.
- (y1 y2) * (x1 x2): creates a graph with four separate scatter plots.

Required Plot Request:

Plot Request	y * x or (y1 y2) * x or y * (x1 x2) or (y1 y2) * (x1 x2)

Options:

ATTRID	=variable	Associated Attr Map for visual attributes
COLUMNS	=value	Number of columns in panel
DATALABEL	=variable	Variable for labeling of scatter points
ELLIPSE	<=(options)	Fit a confidence ellipse to the data
GROUP	= variable	Grouping variable
JOIN	<=options>	Connect the data points
LEGEND	= (options)	Controls legend characteristics
LOESS	= (options)	Fit a Loess curve to the data
MARKERATTRS	=marker-attrs	Specify marker attributes
PBSPLINE	=(options)	Fit a penalized B-Spline to the data
REG	=(options)	Fit a regression line to the data
ROWS	=value	Number of rows in panel
SPACING	=value	Spacing between rows and columns
TRANSPARENCY	=value	Transparency value
UNISCALE	=keyword	Uniform scaling for X	Y	ALL. Default is NONE

Boolean Options:

GRID	Display borders
NOLEGEND	Suppress display of legend
REFTICK	Display duplicate reference tick marks

Figure 11.3.1: Basic Scatter Plots

This is a basic two-cell scatter plot of height by weight and cholesterol by age at start. Each plot is specified as a separate request.

Each cell has its own set of axes with independent data ranges.

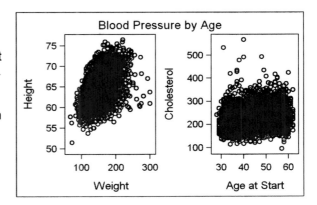

```
title "Blood Pressure by Age";
proc sgscatter data=sashelp.heart;
   plot height*weight cholesterol*ageatstart;
   run;
```

Figure 11.3.2: Scatter Plots with Attributes

This is the same graph as shown in Figure 11.3.1 with the addition of marker attributes and transparency.

The high transparency allows us to see where the observations are clustered.

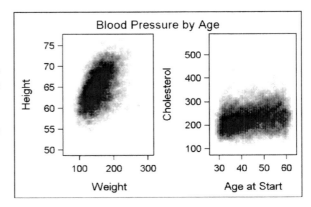

```
title "Blood Pressure by Age";
proc sgscatter data=sashelp.heart;
   plot height*weight cholesterol*ageatstart/
        markerattrs=(symbol=circlefilled) transparency=0.96;
   run;
```

Figure 11.3.3: Plot Grid - 1 Row x 2 Columns

This graph displays diastolic and systolic blood pressure by age. Gridlines have been enabled.

Notice the data range for the two y-axes is not uniform.

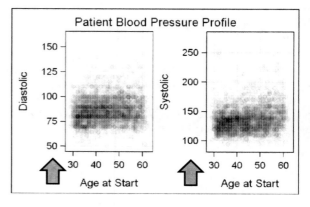

```
title " Patient Blood Pressure Profile";
proc sgscatter data=sashelp.heart;
  where sex='Male';
  plot (diastolic systolic) * ageatstart / grid
       markerattrs=(symbol=circlefilled) transparency=0.96;
run;
```

Figure 11.3.4: Plot Grid - 1 Row x 2 Columns

This graph is the same as in Figure 11.3.3. We have UNISCALE=Y to make all the y-axes have the same scale visually showing the relative values of the measures.

Each cell still has its own x and y-axes.

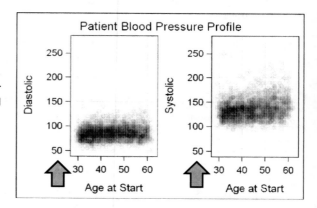

```
title " Patient Blood Pressure Profile";
proc sgscatter data=sashelp.heart;
  where sex='Male';
  plot (diastolic systolic) * ageatstart / grid  uniscale=y
       markerattrs=(symbol=circlefilled) transparency=0.96;
run;
```

Figure 11.3.5: Plot with Fit

This is the same graph as in Figure 11.3.3, with an added regression fit and confidence band.

The transparency of the markers is increased to make the fit line and bands easier to see.

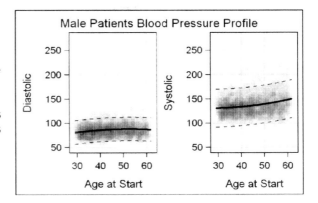

```
title "Male Patients Blood Pressure Profile";
proc sgscatter data=sashelp.heart;
  where sex='Male';
  plot (diastolic systolic) * ageatstart / grid transparency=0.98
       markerattrs=(symbol=circlefilled) uniscale=Y reg=(degree=2 cli);
run;
```

Figure 11.3.6: Plot with Fit and Ellipse

This graph uses the penalized B-spline fit and a confidence ellipse.

The transparency of the markers is increased to make the fit line and ellipse easier to see.

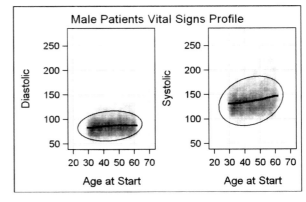

```
title "Male Patients Vital Signs Profile";
proc sgscatter data=sashelp.heart;
  where sex='Male';
  plot (diastolic systolic) * ageatstart / grid
       pbspline=(clm degree=1) ellipse uniscale=Y
       markerattrs=(symbol=circlefilled) transparency=0.98;
run;
```

Figure 11.3.7: Multiple Plot Request

This graph shows diastolic and systolic blood pressure by age and cholesterol by weight in one row per the plot request. Regression fit is shown.

The transparency of the markers is increased to make the fit line and bands easier to see.

```
title "Male Patients Vital Signs
Profile";
proc sgscatter data=sashelp.heart;
  where sex='Male';
  plot (diastolic systolic) *
    ageatstart cholesterol*weight
    / grid Reg=(cli degree=2)
      markerattrs=(symbol=circlefilled)
      transparency=0.98 rows=1;
run;
```

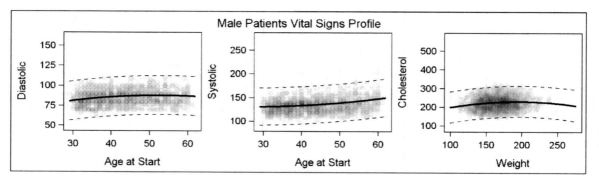

Figure 11.3.8: Multiple Plot Request

This graph shows diastolic and systolic blood pressure and "Cholesterol by Age" and "Age at Death" by "Age at Start". A penalized B-spline fit is shown.

Marker transparency is set to make the fit line and ellipses easier to see.

```
title "Male Patients Vital Signs
Profile";
proc sgscatter data=sashelp.heart;
  where sex='Male';
  plot (diastolic systolic cholesterol)
    * ageatstart ageatdeath * ageatstart
    / grid pbspline=(clm degree=1)
      markerattrs=(symbol=circlefilled)
      transparency=0.98 rows=1 ellipse;
run;
```

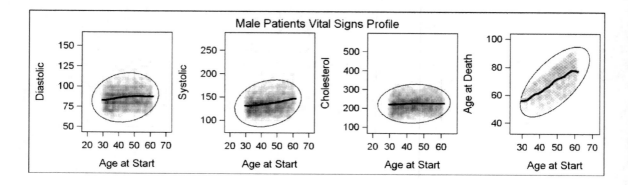

Figure 11.3.9: Uniform Side-by-Side Plots

This is a graph of the response over time for "Drug A" for female and male patients.

We have used the JOIN, UNISCALE and GRID options.

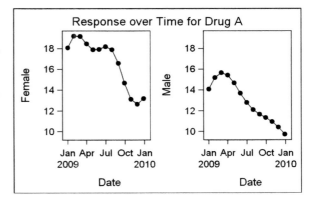

```
title " Response over Time for Drug A";
proc sgscatter data=sgbook.seriesGroup2;
  where drug='A';
  plot (female male)*date / grid uniscale=Y join;
  run;
```

Figure 11.3.10: Uniform Side-by-Side Plots

This is a graph of the response over time by treatment for female and male patients.

We have used the JOIN, UNISCALE and GRID options. Markers are suppressed by setting size=0.

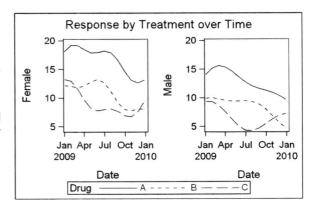

```
title "Response by Treatment over Time";
proc sgscatter data=sgbook.seriesGroup2;
  plot (female male)*date / group=drug grid uniscale=Y join
        markerattrs=(size=0);
  run;
```

11.4 COMPARE Statement

The COMPARE statement creates regular grids of scatter plots. Fit lines and ellipses can be included. The graphs always have common row and column axis. By default, individual columns have uniform data ranges for the x-axis. Individual rows have uniform data range for the y-axis.

```
compare x=(var <var> ...) y=(var <var> ...) </ options>;
```

Required Plot Request:

X	=var-list	List of variables for columns
Y	=var-list	List of variables for rows

Options:

ATTRID	=variable	Associated Attr Map for visual attributes
DATALABEL	=variable	Variable for labeling of scatter points
ELLIPSE	<=(options)	Fit a confidence ellipse to the data
GROUP	= variable	Grouping variable
JOIN	<=options>	Connect the data points
LEGEND	= (options)	Controls legend characteristics
LOESS	= (options)	Fit a Loess curve to the data
MARKERATTRS	=attrs	Specify marker attributes
PBSPLINE	=(options)	Fit a penalized B-spline to the data
REG	=(options)	Fit a regression line to the data
SPACING	=value	Spacing between rows and columns
TRANSPARENCY	=value	Transparency value

Boolean Options:

GRID	Display grid lines
NOLEGEND	Suppress display of legend
REFTICK	Display duplicate reference tick marks

Figure 11.4.1: Comparative Graph

This is a comparative graph of vehicle by city and highway mileage by horsepower. The x-axis is automatically made common, so it is easier to compare the data in the cells with uniform scale.

Gridlines are enabled for ease of comparison. The JournalSmallFont2 style has smaller axis label fonts that are needed for the y-axis labels to fit.

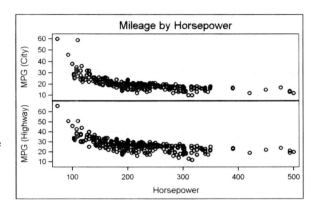

```
title "Mileage by Horsepower";
proc sgscatter data=sashelp.cars;
  compare y=(mpg_city mpg_highway) x= horsepower / grid;
  run;
```

Figure 11.4.2: Comparative Graph

This is a comparative graph showing city mileage by horsepower and weight. Common y-axis reduces clutter and provides more space to display the data.

The x-axis is not uniform. Options are used for filled markers and transparency.

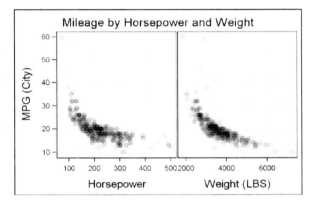

```
title "Mileage by Horsepower and Weight";
proc sgscatter data=sashelp.cars;
  compare y=(mpg_city) x= (horsepower weight) / grid
       markerattrs=(symbol=circlefilled) transparency=0.9;
  run;
```

Figure 11.4.3: Grouped Comparative Plot

This is a comparative graph showing city mileage by horsepower and weight and grouped by country of origin.

Common y-axis reduces clutter and provides more space to display the data. The x-axis is not uniform.

Options are used for transparency and marker size.

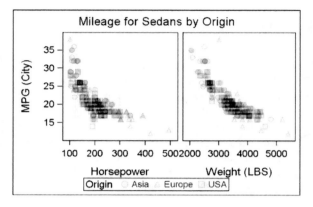

```
title "Mileage for Sedans by Origin";
proc sgscatter data=sashelp.cars;
  where type='Sedan';
  compare y=(mpg_city) x= (horsepower weight) / group=origin grid
          transparency=0.9 markerattrs=(size=9);
  run;
```

Figure 11.4.4: 2x2 Grouped ComparativeGraph

This is a comparative plot of city and highway mileage by horsepower and weight. Common row and column axes are used by default, so individual row and column ranges are uniform.

The quadratic regression fit is not by group (NOGROUP).

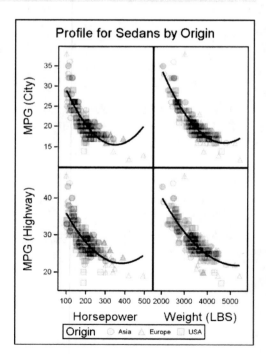

```
title "Profile for Sedans by Origin";
proc sgscatter data=sashelp.cars;
  where type='Sedan';
  compare y=(mpg_city mpg_highway)
          x= (horsepower weight) /
          group=origin grid
          transparency=0.9
          markerattrs=(size=9)
          reg=(nogroup degree=2);
  run;
```

Figure 11.4.5: Comparative Graph

This is a comparative graph of male and female response over time for "Drug A". Since two Y variables and one X variable are provided, we get a single column with two rows.

Common x-axis is displayed with grids.

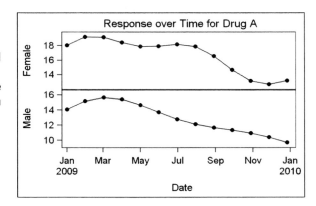

```
title " Response over Time for Drug A";
proc sgscatter data=sgbook.seriesGroup2;
  where drug='A';
  compare y=(female male) x=date / grid join;
  run;
```

Figure 11.4.6: Comparative Graph

This is a comparative graph of systolic blood pressure by weight and age grouped by sex for patients who died of coronary heart failure.

Linear regression fit lines by sex are displayed. Transparency is used to unclutter the display.

Common y-axis is displayed with grids.

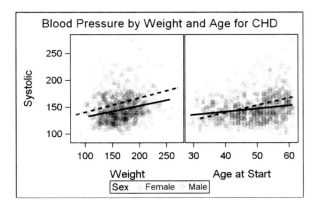

```
title "Blood Pressure by Weight and Age for CHD";
proc sgscatter data=sashelp.heart;
  where deathcause = 'Coronary Heart Disease';
  compare y=systolic x=(weight ageatstart)  / group=sex
        grid reg transparency=0.9;
  run;
```

Figure 11.4.7: Heat Map

This is heat map showing the frequency of occurrences for patients who died of coronary heart disease (CHD) by cholesterol status, blood pressure status, sex, and weight status.

Darker squares represent higher occurrence. This graph shows highest levels of CHD deaths for overweight males with high blood pressure and cholesterol.

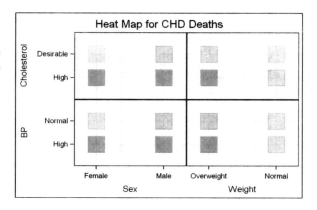

```
title "Heat Map for CHD Deaths";
proc sgscatter data=sgbook.HeatMap;
  label chol_status='Cholesterol';
  label bp_status='BP';
  label weight_status='Weight';
  compare y=(chol_Status bp_Status) x=(sex Weight_Status)  / grid
          transparency=0.995 markerattrs=(symbol=squarefilled size=30);
  run;
```

Figure 11.4.8: Rectangular Matrix

This is a rectangular matrix of scatter plots with confidence ellipse showing the relationship of age and weight by systolic and diastolic blood pressure and cholesterol for CHD patients.

Some correlation between weight and blood pressure can be visually discerned.

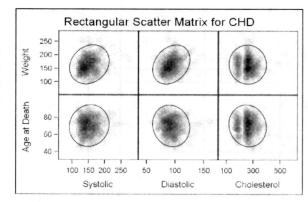

```
title "Rectangular Scatter Matrix for CHD";
proc sgscatter data=sgbook.HeatMap;
  compare x=(systolic diastolic cholesterol) y=(weight ageatdeath)   /
          markerattrs=(symbol=circlefilled size=10) ellipse
          grid transparency=0.95;
  run;
```

11.5 MATRIX Statement

The MATRIX statement creates regular grids of scatter plots of pair-wise combinations of all the variables in the list. Diagonals are used to label the variable name. Diagonals can alternatively be used to display the distribution of each individual variable itself:

```
matrix var var <var> <var> ... </ options>;
```

Required Plot Request:
    Plot Request   Var list

Options:

ATTRID	=variable	Associated Attr Map for visual attributes
DATALABEL	=variable	Variable for labeling of scatter points
ELLIPSE	<=(options)	Fit a confidence ellipse to the data
GROUP	=variable	Grouping variable
JOIN	<=options>	Connect the data points
LEGEND	=(options)	Controls legend characteristics
MARKERATTRS	=attrs	Specify marker attributes
START	=keyword	Specify the starting position – TOPLEFT | BOTTOMLEFT
TRANSPARENCY	=value	Specify the transparency for the markers

Boolean Options:

NOLEGEND		Suppress display of legend

Figure 11.5.1: Basic Scatter Plot Matrix

This is a scatter plot matrix of "Weight" x "Cholesterol" x "Systolic" for female patients with coronary heart disease.

A scatter plot is drawn for each pair-wise combination of the variables specified. The diagonals display the variable name.

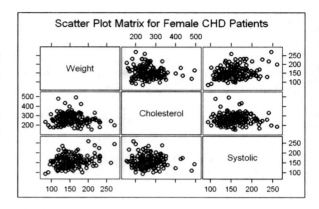

```
title "Scatter Plot Matrix for CHD Patients";
proc sgscatter data=sashelp.heart;
  where sex='Female';
  matrix weight cholesterol systolic;
  run;
```

Figure 11.5.2: Basic Scatter Plot Matrix

This is a scatter plot matrix of "Weight" x "Cholesterol" x "Systolic" for female patients with coronary heart disease.

A filled marker is used with a transparency of 80%. This provides us a better view of where the data is clustered.

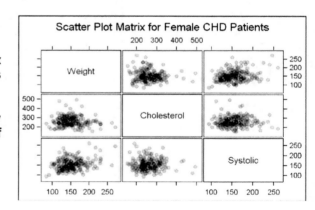

```
title "Scatter Plot Matrix for Female CHD Patients";
proc sgscatter data=sgbook.chd;
  where sex='Female';
  matrix weight cholesterol systolic / transparency=0.8
         markerattrs=(symbol=circlefilled);
  run;
```

Figure 11.5.3: Scatter Plot Matrix with Histograms

This is a scatter plot matrix of "Weight" x "Cholesterol" x "Systolic" with a histogram of each individual variable shown along the diagonal.

Variable labels are moved out to the top and side.

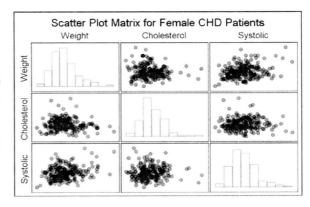

```
title "Scatter Plot Matrix for Female CHD Patients";
proc sgscatter data=sgbook.chd;
   where sex='Female';
   matrix weight cholesterol systolic / transparency=0.6
          markerattrs=(symbol=circlefilled) diagonal=(histogram);
   run;
```

Figure 11.5.4: Scatter Plot Matrix with Diagonals and Ellipse

This is a scatter plot matrix of "Weight" x "Cholesterol" x "Systolic" with histogram and density plots of each individual variable shown along the diagonal.

A confidence ellipse is shown for each scatter plot.

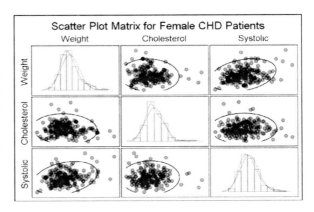

```
title "Scatter Plot Matrix for Female CHD Patients";
proc sgscatter data=sgbook.chd;
   where sex='Female';
   matrix weight cholesterol systolic / transparency=0.6 ellipse
          markerattrs=(symbol=circlefilled)
          diagonal=(histogram normal kernel);
   run;
```

Figure 11.5.5: Grouped Scatter Plot Matrix

This is a scatter plot matrix of "MPG" x "Horsepower" x "Weight". Country of origin is set as the group variable.

Ellipses are clipped to fit the data, and a legend is displayed.

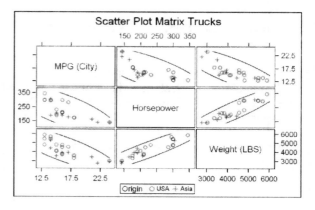

```
title "Scatter Plot Matrix Trucks";
proc sgscatter data=sashelp.cars;
  where type='Truck';
  matrix mpg_city horsepower weight / group=origin  ellipse
        transparency=0.5;
  run;
```

Figure 11.5.6: 4x4 Scatter Plot Matrix

This is a 4x4 scatter plot matrix of "MPG" x "Horsepower" x "Weight" x "MSRP".

The START option is set to BOTTOMLEFT, so the matrix is laid out bottom to top. The previous examples all had the matrix laid out from top down.

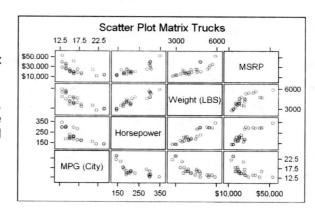

```
title "Scatter Plot Matrix Trucks";
proc sgscatter data=sashelp.cars;
  where type='Truck';
  matrix mpg_city horsepower weight msrp/ start=bottomleft
        transparency=0.5;
  run;
```

Chapter 12

Health and Life Sciences Graphs

12.1 Introduction   287
12.2 Forest Plot   288
12.3 Forest Plot (9.3)   289
12.4 Survival Plot   290
12.5 Adverse Event Timeline   291
12.6 Adverse Event Timeline (9.3)   292
12.7 Maximum LFT Values by Treatment (9.3)   293
12.8 Median of Lipid Profile over Time by Treatment (9.3)   294
12.9 QTc Change from Baseline over Time by Treatment (9.3)   295
12.10 QTc Change Graph with Annotated "At Risk" Values (9.3)   296
12.11 QTc Change from Baseline over Time by Treatment   297
12.12 LFT Safety Panel, Baseline vs. Study   298
12.13 Immunology Profile by Treatment   299
12.14 Most Frequent On-Therapy Adverse Events by Frequency   300
12.15 LFT Patient Profile   301
12.16 Panel of LFT Values   302
12.17 Distribution of Eye Irritation Using PROC SGPANEL (9.3)   303
12.18 Distribution of Eye Irritation Using PROC SGPLOT (9.3)   304
12.19 Vital Signs by Time Point Name   305
12.20 Concomitant Medications   306
12.21 Creating a 2 x 2 Cell Graph Using PROC SGPLOT   307

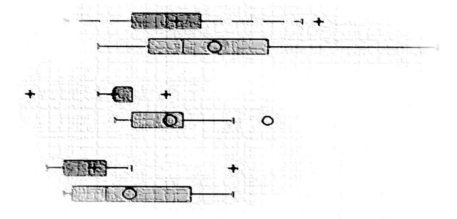

"We didn't lose the game; we just ran out of time."
~ Vince Lombardi

Chapter 12: Health and Life Sciences Graphs

12.1 Introduction

The health and life sciences industry is an ideal candidate for use of graphs. Data is collected for many different applications such as clinical trials, demographics, vital signs, laboratory test results for chemistry and hematology, patient profiles, and adverse events.   Such data is initially collected in tabular form, and often consumed in the same form.

Evaluation of such data can be significantly enhanced by the use of graphical displays of the data in the form of dot plots, box plots, and lattice and matrix displays. Graphical display of data provides insights into trends and correlations that are just not possible through tabular data. Often, combining the data and summary statistics on a single display allows for improved analysis and interpretation of the results.

The examples include displays of lab results over time, distribution of tests by treatment, lattice and matrix displays of liver function tests (LFT), patient profiles, adverse event plots sorted by relative risk, hazard function plots, and displays for evaluation of blood chemistry and hematology.

SG procedures are very well suited for the creation of such displays using the techniques described in the previous chapters of this book. In this chapter, we will examine how you can create some common graphs used in this industry. Some examples are shown in Figure 12.1. Detailed code for creating such graphs is shown in the following pages.

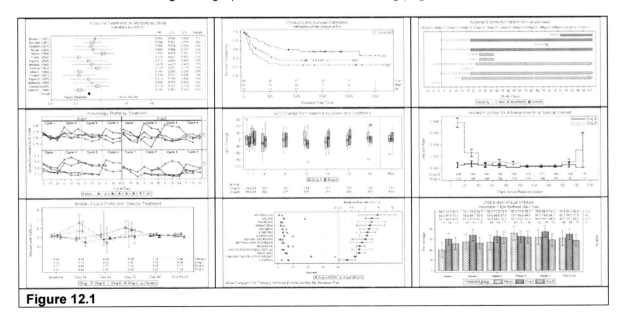

Figure 12.1

12.2 Forest Plot

This graph shows a common version of the forest plot, showing the results from different studies and an overall observation. The data is as shown in the table below. The key features are:
- Plot the graph on the X-Y axis (numeric), restricted to the lower 65% of the x-axis.
- Plot the data on the X2-Y axis (character), restricted to the upper 30% of the X2 axis.
- Use multiple scatter plots with MARKERCHAR option to display the statistics.

Study	OddsRatio	LowerCL	UpperCL	Weight	Q1	Q3	ObsId	study2	lcl2	ucl2	OR	LCL	UCL	WT
Modano (1967)	0.590	0.096	3.634	5%	0.56	0.62	1		0.096	3.634	OR	LCL	UCL	Weight
Borodan (1981)	0.464	0.201	1.074	18%	0.38	0.55	2		0.201	1.074	OR	LCL	UCL	Weight
	0.328	0.233	0.462		.	.	.	16	Overall	.	.	OR	LCL	UCL	Weight

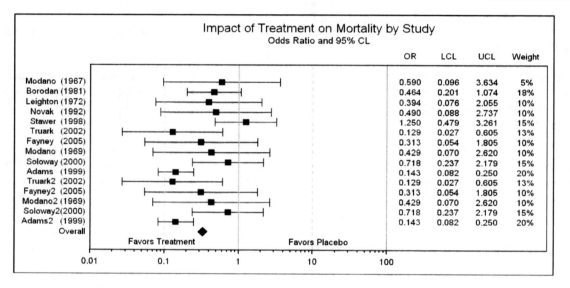

```
title "Impact of Treatment on Mortality by Study";
title2 h=8pt 'Odds Ratio and 95% CL';
proc sgplot data=sgbook.forest noautolegend;
  scatter y=study2 x=oddsratio / markerattrs=(symbol=diamondfilled size=10);
  scatter y=study x=oddsratio / xerrorupper=ucl2 xerrorlower=lcl2
       markerattrs=(symbol=squarefilled);
  scatter y=study x=or / markerchar=oddsratio x2axis;
  scatter y=study x=lcl / markerchar=lowercl x2axis;
  scatter y=study x=ucl / markerchar=uppercl x2axis;
  scatter y=study x=wt / markerchar=weight x2axis;
  refline 1 100 / axis=x;
  refline 0.01 0.1 10 / axis=x lineattrs=(pattern=shortdash) transparency=0.5;
  inset '            Favors Treatment' / position=bottomleft;
  inset 'Favors Placebo'  / position=bottom;
  xaxis type=log offsetmin=0 offsetmax=0.35 min=0.01 max=100 minor display=(nolabel) ;
  x2axis offsetmin=0.7 display=(noticks nolabel);
  yaxis display=(noticks nolabel) offsetmin=0.1 offsetmax=0.1;
run;
```

12.3 Forest Plot (9.3)

This graph shows the same forest plot from Figure 12.2 using the HighLow plot available with SAS 9.3. The data is as shown below. The key features of this plot are:
- A HighLow plot with TYPE=Bar is used to display of the weights of the study.
- A HighLow plot with TYPE=Line (Default) is used to display the confidence limits.

Study	OddsRatio	LowerCL	UpperCL	Weight	Q1	Q3	ObsId	study2	lcl2	ucl2	OR	LCL	UCL	WT
Modano (1967)	0.590	0.096	3.634	5%	0.56	0.62	1		0.096	3.634	OR	LCL	UCL	Weight
Borodan (1981)	0.464	0.201	1.074	18%	0.38	0.55	2		0.201	1.074	OR	LCL	UCL	Weight
	0.328	0.233	0.462		.	.	16	Overall	.	.	OR	LCL	UCL	Weight

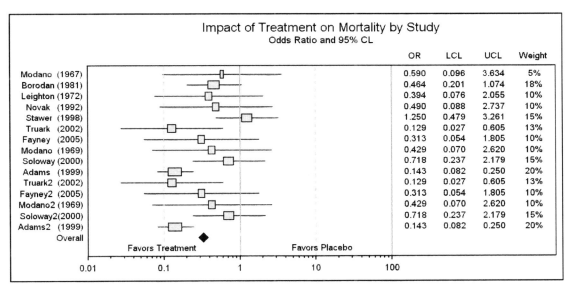

```
title "Impact of Treatment on Mortality by Study";
title2 h=8pt 'Odds Ratio and 95% CL';
proc sgplot data=sgbook.forest2 noautolegend nocycleattrs;
  scatter y=study2 x=oddsratio / markerattrs=(symbol=diamondfilled);
  highlow y=study low=lcl2 high=ucl2 / type=line;
  highlow y=study low=q1 high=q3 / type=bar;
  scatter y=study x=or / markerchar=oddsratio x2axis;
  scatter y=study x=lcl / markerchar=lowercl x2axis;
  scatter y=study x=ucl / markerchar=uppercl x2axis;
  scatter y=study x=wt / markerchar=weight x2axis;
  refline 1 100     / axis=x;
  refline 0.01 0.1 10 / axis=x lineattrs=(pattern=shortdash) transparency=0.5;
  inset '          Favors Treatment'  / position=bottomleft;
  inset 'Favors Placebo'   / position=bottom;
  xaxis type=log offsetmin=0 offsetmax=0.35 min=0.01 max=100 minor display=(nolabel) ;
  x2axis offsetmin=0.7 display=(noticks nolabel);
  yaxis display=(noticks nolabel) offsetmin=0.1 offsetmax=0.2 reverse;
run;
```

12.4 Survival Plot

This graph shows the survival estimates for multiple groups of patients by severity of disease. The data is obtained using the "ods output" from the "SurvivalPlot" object of Sample 49.2.1 of the LIFETEST procedure as shown below. The key features of the graph are:
- Plot survival curves on the X-Y axis (numeric), restricted to the upper 75% of the Y axis.
- Plot at risk data on the X-Y2 axis (character), restricted to the lower 17% of the Y2 axis.
- Survival curves are labeled using curve labels for clarity. Legend works for color style.

Obs	Time	Survival	AtRisk	Event	Censored	tAtRisk	Stratum	StratumNum
1	0	1.00000	38	0	.	.	1: ALL	1
2	0	.	38	.	.	0	1: ALL	1

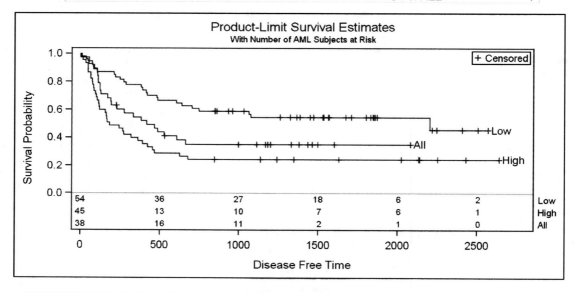

```
ods output Survivalplot=sgbook.SurvivalPlot49_2_1;
proc lifetest data=BMT plots=survival(atrisk=0 to 2500 by 500);
   time T * Status(0);
   strata Group / test=logrank adjust=sidak; run;
proc format;   value aml   3 = 'Low'   2 = 'High'   1 = 'All'; run;

title 'Product-Limit Survival Estimates';
title2 h=7pt 'With Number of AML Subjects at Risk';
proc sgplot data=sgbook.SurvivalPlot49_2_1;
  format stratumNum aml.;
  step x=time y=survival / group=stratumNum curvelabel lineattrs=(pattern=solid);
  scatter x=time y=censored / markerattrs=(symbol=plus) name='censored';
  scatter x=tatrisk y=stratumnum / markerchar=atrisk y2axis group=stratumnum;
  keylegend 'censored' / location=inside position=topright;
  refline 0;
  yaxis offsetmin=0.2 min=0;
  y2axis offsetmax=0.83 display=(nolabel noticks) valueattrs=(size=8);
  run;
```

12.5 Adverse Event Timeline

This graph shows adverse events by time and severity. The data is extracted from Clinical Data Interchange Standards Consortium (CDISC) data and is as shown in the table below. The key features of the graph are:
- Vector plots are used to display the events.
- Scatter plots are used to show the start and end events.
- Scatter plot with MARKERCHAR option is used to display the event name.
- Note: X position (xc) for the label needs to be computed. See detailed program.

Obs	aestdate	aeendate	aeseq	aedecod	aesev	y	xc	aestdy	aeendy	startday	endday
1	06MAR13	06MAR13	1	DUMMY	MILD	-100	-4.4	0	0	0	0
2	06MAR13	06MAR13	1	DUMMY	MODERATE	-100	-4.4	0	0	0	0
3	06MAR13	06MAR13	1	DUMMY	SEVERE	-100	-4.4	0	0	0	0
4	06MAR13	06MAR13	1	DIZZINESS	MODERATE	1	-6.6	0	0	0	0
5	20MAR13	.	2	COUGH	MILD	2	9.57	14	.	14	104

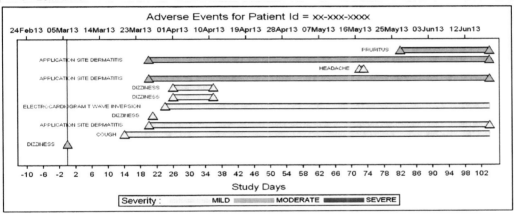

```
title "Adverse Events for Patient Id = xx-xxx-xxxx";
proc sgplot data=sgbook.ae1 noautolegend nocycleattrs;
    refline 0 / axis=x lineattrs=(thickness=1 color=black);
    vector x=endday y=y / xorigin=startday yorigin=y noarrowheads
           lineattrs=(thickness=9px);
    vector x=endday y=y / xorigin=startday yorigin=y group=aesev noarrowheads
           lineattrs=(thickness=7px pattern=solid) transparency=0 name='sev';
    scatter x=aestdy y=y / markerattrs=(size=13px symbol=trianglefilled);
    scatter x=aestdy y=y / markerattrs=(size=9px symbol=trianglefilled) group=aesev;
    scatter x=aeendy y=y / markerattrs=(size=13px symbol=trianglefilled);
    scatter x=aeendy y=y / markerattrs=(size=9px symbol=trianglefilled) group=aesev;
    scatter x=xc y=y / markerchar=aedecod;
    scatter x=aestdate y=y / markerattrs=(size=0) x2axis;
    yaxis display=(nolabel noticks novalues) min=0;
    xaxis grid label='Study Days' offsetmin=0.02 offsetmax=0.02
          values=(&minday10 to &maxday by 2);
    x2axis notimesplit display=(nolabel) offsetmin=0.02 offsetmax=0.02
          values=(&mindate10 to &maxdate);
    keylegend 'sev'/ title='Severity :';
run;
```

12.6 Adverse Event Timeline (9.3)

This graph shows adverse events by time and severity. The data is extracted from CDISC data and is as shown in the table below. The key features for the graph are:
- The HighLow option is used to display the events.
- The LOWLABEL option is used to display the event name.
- Low and high caps are used to display events that are continuing.

aestdate	aeendate	aeseq	aedecod	aesev	y	aestdy	aeendy	stday	enday	lcap	hcap	xs
06MAR13	06MAR13	.		MILD	-9	0	0	0	0			0
06MAR13	06MAR13	.		MODERATE	-9	0	0	0	0			0
06MAR13	06MAR13	.		SEVERE	-9	0	0	0	0			0
06MAR13	06MAR13	1	DIZZINESS	MODERATE	1	0	0	0	0			0
20MAR13	.	2	COUGH	MILD	2	14	.	14	104		ARROW	0

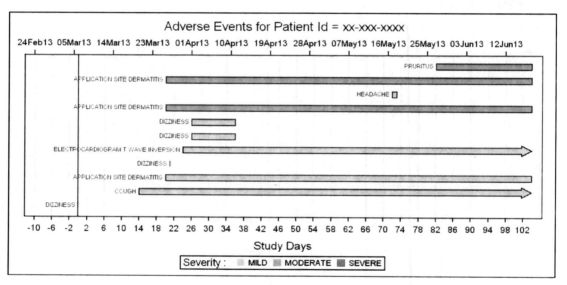

```
title "Adverse Events for Patient Id = xx-xxx-xxxx";
proc sgplot data=sgbook.ae2 noautolegend nocycleattrs;
  highlow y=y low=xs high=xs / group=aesev type=bar barwidth=0.8 name='sev' Y2axis;
  highlow y=aeseq low=stday high=enday / group=aesev lowlabel=aedecod type=bar
      barwidth=0.8 lowcap=lcap highcap=hcap lineattrs=(pattern=solid color=black);
  scatter x=aestdate y=aeseq /  markerattrs=(size=0) x2axis;
  refline 0 / axis=x lineattrs=(thickness=1 color=black);
  yaxis display=(nolabel noticks novalues) type=discrete;
  y2axis display=none min=0;
  xaxis grid label='Study Days' offsetmin=0.02 offsetmax=0.02
       values=(&minday10 to &maxday by 2);
  x2axis notimesplit display=(nolabel) offsetmin=0.02 offsetmax=0.02
       values=(&mindate10 to &maxdate);
  keylegend 'sev'/ title='Severity :';    run;
```

12.7 Maximum LFT Values by Treatment (9.3)

This graph shows the "Distribution of Maximum Liver Function Test Values by Treatment" using a grouped box plot on a discrete axis. The data is as shown on the right.
- Use a box plot grouped by "Drug".
- Use a reference line to display the clinical concern levels.

Obs	Test	Drug	Value
1	ALAT	A	1.05198
2	ALAT	B	0.97755
3	ASAT	A	0.78177
4	ASAT	B	0.59554
5	ALKPH	A	0.20475

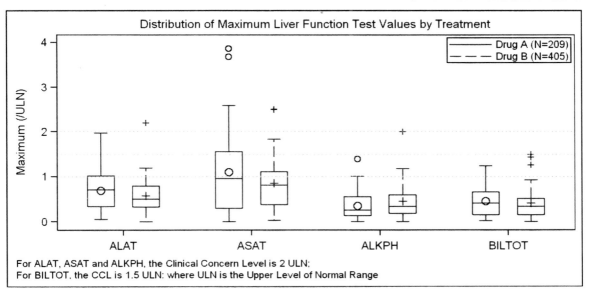

```
title h=10pt 'Distribution of Maximum Liver Function Test Values ...';
footnote1 h=8pt j=left "For ALAT, ASAT and ALKPH, the CCL is 2 ULN;";
footnote2 h=8pt j=left "For BILTOT, the CCL is 1.5 ULN: where ULN is the Upper Level of
Normal Range";
proc sgplot data=sgbook.LFT_Group;
  format drug $drug.;
  vbox value / category=test group=drug nofill lineattrs=(pattern=solid)
          medianattrs=(pattern=solid);
  keylegend / location=inside position=topright across=1;
  refline 1 1.5 2 / lineattrs=(pattern=dot);
  xaxis discreteorder=data display=(nolabel);
  yaxis label='Maximum (/ULN)';
run;
```

12.8 Median of Lipid Profile over Time by Treatment (9.3)

In this graph we are displaying the lipid profile over time by treatment group. Here again, we have used the "Y-Y2 Split" technique to split the graph region into two parts:

X	Drug	Median	Lcl	Ucl
Baseline	Drug A	5.21	5.04	5.52
Baseline	Drug B	5.17	4.94	5.47
Baseline	Drug C	5.24	4.97	5.33
Baseline	Placebo	5.08	4.81	5.35
Day 14	Drug A	4.90	4.60	5.79
Day 14	Drug B	6.65	4.81	7.51

- The upper region is used to display the graph.
- The lower region is used to display the statistics.
- A cluster grouped scatter plot is used for the values.
- A cluster grouped series plot is used for the lines.
- The cluster width is reduced to tighten the clusters.
- A scatter plot with y=Drug and MARKERCHAR option is used to display the statistics.
- A reference line on the Y2 axis, with DISCRETEOFFSET, is used to separate the two regions.

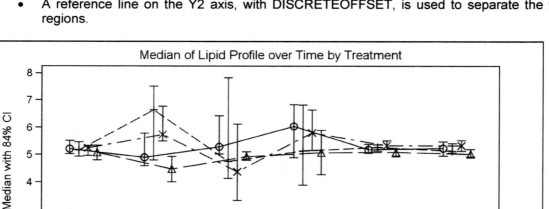

```
title 'Median of Lipid Profile over Time by Treatment';
proc sgplot data=sgbook.lipid_grp;
  where by='Test 1';
  scatter x=x y=median / group=drug yerrorupper=ucl yerrorlower=lcl
          groupdisplay=cluster clusterwidth=0.5 markerattrs=(size=9);
  series x=x y=median / group=drug groupdisplay=cluster clusterwidth=0.5;
  scatter x=x y=drug / markerchar=median y2axis;
  refline 'Placebo' / discreteoffset=0.5 axis=y2;
  xaxis display=(nolabel);
  yaxis grid offsetmin=0.30;
  y2axis offsetmax=0.8 display=(noticks nolabel) valueattrs=(size=7);
run;
```

12.9 QTc Change from Baseline over Time by Treatment (9.3)

This graph displays the change in corrected QT (QTc) interval over time for heartbeats by treatment. The visits are irregularly spaced, with higher frequency in the beginning, so it is desirable to have a linear axis for time. This graph has the following features:

Week	Week2	atRiskA	atRiskB	valueA	valueB
1	0.75	209	.	72.1929	.
1	0.75	209	.	-3.1966	.
1	1.25	.	405	.	-29.5119
1	1.25	.	405	.	-10.0056
2	1.75	206	.	52.2286	.
2	1.75	206	.	57.0051	.

- An overlay of two VBOX statements is used for the data.
- The x-axis values are jittered by 0.25 for each treatment to get adjacent overlays.
- A user-defined format is used to label the last value as "Max".
- A reference line is added to separate the last value.
- Ideally, we would like to display the "atRisk" values for each week of the study. To do so, we will use the Annotation feature for SGPLOT as shown in Figure 12.10.

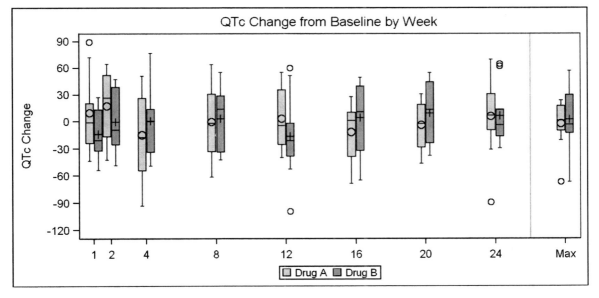

```
title 'QTc Change from Baseline by Week';
proc sgplot data=sgbook.QtcChange;
   vbox valueA / category=Week2 lineattrs=(pattern=solid) meanattrs=graphdata1
           whiskerattrs=(pattern=solid) medianattrs=(pattern=solid);
   vbox valueB / category=Week2 lineattrs=(pattern=solid) meanattrs=graphdata2
           whiskerattrs=(pattern=solid) medianattrs=(pattern=solid);
   refline 26 / axis=x;
   xaxis type=linear values=(1 2 4 8 12 16 20 24 28) display=(nolabel)
           tickvalueformat=myWeek. offsetmin=0.03 offsetmax=0.03;
   yaxis values=(-120 to 90 by 30) grid label='QTc Change';
run;
```

12.10 QTc Change Graph with Annotated "At Risk" Values (9.3)

Here we have added the "At Risk" table to the graph shown in Figure 12.9 using the Annotation.
- An annotation data set is created to draw the at-risk values for "Drug A" and "Drug B".
- The X coordinate is in DataValue space so they line up with the X axis values.
- The Y coordinate is in WallPixel space, 62 and 74 pixels **below** the y-axis.
- Annotations are also created for the labels and the "At Risk" label.
- Space is created for the annotation by adding a bottom pad of 80 pixels.

X1Space	Y1Space	Label	Function	Anchor	TextSize	X1	Y1
DataValue	WallPixel	209	Text	Center	8	1	-62
DataValue	WallPixel	405	Text	Center	8	1	-74
WallPercent	WallPixel	At Risk :	Text	Right	8	-1	-50
WallPercent	WallPixel	Drug A :	Text	Right	8	-1	-62
WallPercent	WallPixel	Drug B :	Text	Right	8	-1	-74

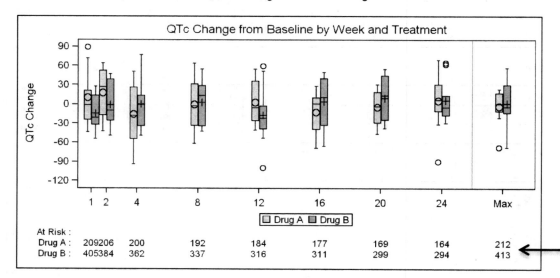

```
title 'QTc Change from Baseline by Week';
proc sgplot data=sgbook.QtcChange sganno=sgbook.QTcAnno pad=(bottom=80);
  vbox valueA / category=Week2 lineattrs=(pattern=solid) meanattrs=graphdata1
          whiskerattrs=(pattern=solid) medianattrs=(pattern=solid);
  vbox valueB / category=Week2 lineattrs=(pattern=solid) meanattrs=graphdata2
          whiskerattrs=(pattern=solid) medianattrs=(pattern=solid);
  refline 26 / axis=x;
  xaxis type=linear values=(1 2 4 8 12 16 20 24 28) display=(nolabel)
        tickvalueformat=myWeek. offsetmin=0.03 offsetmax=0.03;
  yaxis values=(-120 to 90 by 30) grid label='QTc Change';
run;
```

12.11 QTc Change from Baseline over Time by Treatment

This graph displays the hazard function over time by treatment. The features of the graph are:

day	value	Drug	Low	High	Risk	DrugLabel
10	.0005	Drug A	.0001	.0009	212	
10	.0078	Drug B	.0070	.0086	428	
190	.0030	Drug B	.0000	.0060	19	Drug A
190	.0030	Drug B	.0000	.0060	19	Drug B

- The Y-Y2 Split technique is used to display the axis aligned at risk table at the bottom.
- We could have displayed the Y2 labels, but that takes up extra space. We want all labels on one side.
- So, we used the REFLINE statement, with "DrugLabel" as the variable. This has only two values, and these are displayed. The lines are suppressed by setting thickness=0.
- The separator refline uses a value=-0.0005. This is data dependent. It is better to use the refline from the Y2 axis value='Drug B' and DiscreteOffset=0.5, a SAS 9.3 feature.

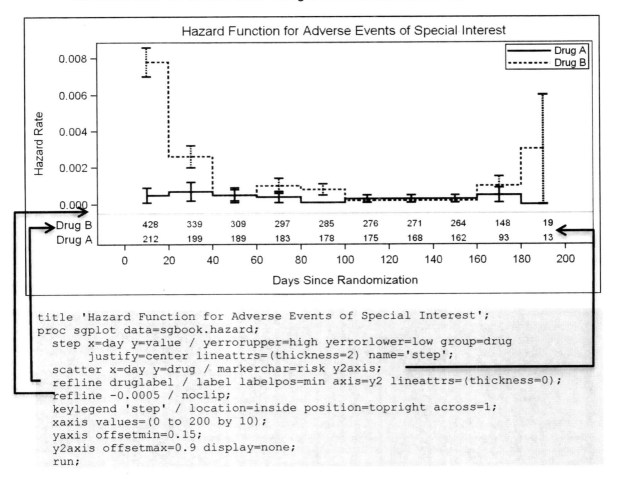

```
title 'Hazard Function for Adverse Events of Special Interest';
proc sgplot data=sgbook.hazard;
  step x=day y=value / yerrorupper=high yerrorlower=low group=drug
       justify=center lineattrs=(thickness=2) name='step';
  scatter x=day y=drug / markerchar=risk y2axis;
  refline druglabel / label labelpos=min axis=y2 lineattrs=(thickness=0);
  refline -0.0005 / noclip;
  keylegend 'step' / location=inside position=topright across=1;
  xaxis values=(0 to 200 by 10);
  yaxis offsetmin=0.15;
  y2axis offsetmax=0.9 display=none;
run;
```

12.12 LFT Safety Panel, Baseline vs. Study

This graph shows the distribution of LFT values by baseline classified by test and time. The data is as shown in the table.

visitnum	labtest	drug	pre	result
1 Week	ALAT	Drug A (n=240)	0.27216	0.64893
1 Week	Bilirubin Total	Drug A (n=240)	1.28536	0.54600
1 Week	Alk Phosphatase	Drug A (n=240)	0.21586	0.01842
1 Week	ASAT	Drug A (n=240)	0.46319	0.64387
3 Months	Alk Phosphatase	Drug B (n=195)	1.67485	0.18387
3 Months	ASAT	Drug B (n=195)	0.07650	0.21848

- Here we use the SGPANEL procedure with Type=Lattice.
- Column class variable is Labtest and Row class variable is Visit.
- Each cell has a scatter plot of Result x Baseline using filled markers with transparency.
- Reflines for multiple clinical concern levels are displayed.

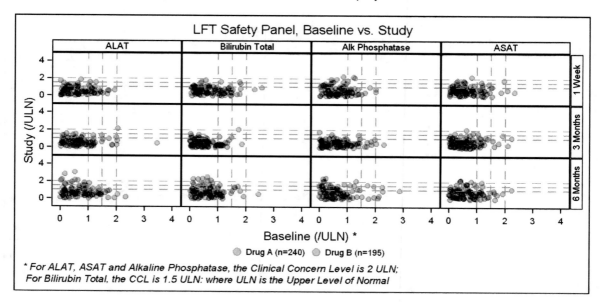

```
title 'LFT Safety Panel, Baseline vs. Study';
footnote j=l italic height=8pt "* For ALAT, ASAT and Alkaline Phosphatase, the...";
footnote2 j=l italic height=8pt " For Bilirubin Total, the CCL is 1.5 ULN: where ...";
proc sgpanel data=sgbook.labs;
panelby labtest visitnum / layout=lattice onepanel novarname;
scatter x=pre y=result/ group=drug markerattrs=(symbol=circlefilled size=9)
        transparency=0.7;
  refline 1 1.5 2 / axis=Y lineattrs=(pattern=dash);
  refline 1 1.5 2 / axis=X lineattrs=(pattern=dash);
  rowaxis integer min=0 max=4 label='Study (/ULN)';
  colaxis integer min=0 max=4 label='Baseline (/ULN) *';
  keylegend/title=" " noborder;
run;
```

12.13 Immunology Profile by Treatment

This graph shows the immunology profile for patients in a study over time by treatment and lab test. The study period is in four cycles.

sival	trt	cyc	pt	lbparm	xval	labelX	labelY
1.120	Drug A	Cycle 1	1	C3	0	2	1.60
0.147	Drug A	Cycle 1	1	C4	0	2	0.45
1.440	Drug A	Cycle 2	1	C3	0	5	1.60
0.278	Drug A	Cycle 2	1	C4	0	5	0.45
1.190	Drug A	Cycle 3	1	C3	0	8	1.60
0.282	Drug A	Cycle 3	1	C4	0	8	0.45
0.917	Drug A	Cycle 4	1	C3	0	11	1.60
0.163	Drug A	Cycle 4	1	C4	0	11	0.45
1.130	Drug B	Cycle 1	5	C3	0	2	1.60
0.220	Drug B	Cycle 1	5	C4	0	2	0.45

- This is created using SGPANEL procedure with TYPE=Lattice.
- The column var is treatment and row var is lbparm.
- The "Cycle" labels at the top are displayed using a scatter plot with MarkerChar. The positions of the labels are set in the data as LabelX and LabelY.
- The 12 observations on the x-axis across four cycles use a format to display the repeating 0, 15, 30 axis tick values.

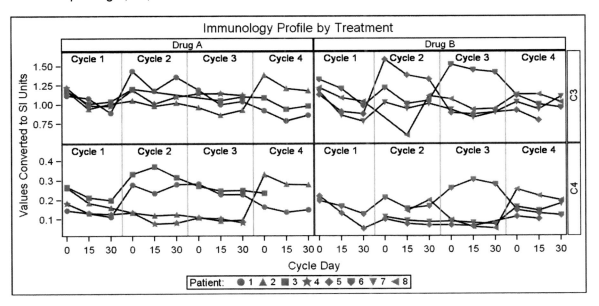

```
title 'Immunology Profile by Treatment';
proc sgpanel data=sgbook.immune;
  panelby trt lbparm / layout=lattice uniscale=column novarname;
  series x=xval y=sival / group=pt lineattrs=(pattern=solid thickness=2);
  scatter x=xval y=sival / group=pt markerattrs=(size=9 color=cx4f4f4f) name='a';
  scatter x=labelx y=labely / markerchar=cyc markercharattrs=(size=9);
  refline 3.5 6.5 9.5 / axis=x;
  keylegend 'a' / title='Patient: ';
  colaxis values=(1 to 12 by 1) offsetmin=0.02 offsetmax=0.02 label='Cycle Day' grid;
  rowaxis grid;
run;
```

12.14 Most Frequent On-Therapy Adverse Events by Frequency

This graph shows the most frequent adverse events for the data shown below:
- The graph is split into two regions using the X-X2 split technique discussed earlier.
- The left side has a plot of percent by adverse event (AE) and treatment on x-axis with OffsetMax=0.5.
- The right side has a plot of the relative risk and CL on X2 axis with OffsetMin=0.5.
- A wide, soft refline is used for the category bands.

AE	A	B	Low	Mean	High	RefAE
DYSPNEA	7.0	2.5	0.13	0.30	0.7	DYSPNEA
CHRONIC OBSTRUCTIVE AIRWAY	22.0	38.0	0.60	0.70	0.8	
BACK PAIN	5.0	6.0	0.80	1.04	2.0	BACK PAIN
GASTROESOPHAGEAL REFLUX	3.0	4.0	0.50	1.05	3.8	
HEADACHE	7.0	10.0	0.80	1.10	2.0	HEADACHE

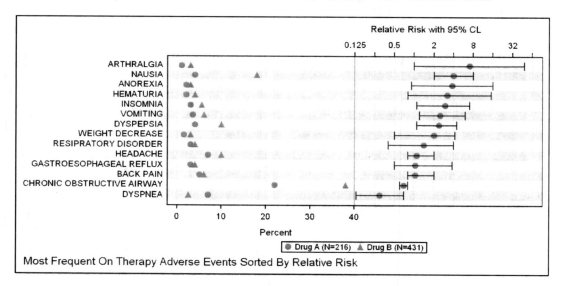

```
footnote j=l 'Most Frequent On Therapy Adverse Events Sorted By Relative Risk';
proc sgplot data=sgbook.MostFrequentAE;
  refline refae / lineattrs=(thickness=12) transparency=0.8;
  scatter y=ae x=a / name='a' legendlabel='Drug A (N=216)';
  scatter y=ae x=b / name='b' legendlabel='Drug B (N=431)';
  scatter y=ae x=mean / xerrorlower=low xerrorupper=high  x2axis;
  refline 40 / axis=x;
  keylegend 'a' 'b';
  xaxis offsetmax=0.5 grid labelattrs=(size=8) valueattrs=(size=7)
        label='Percent                                                                    ';
  x2axis offsetmin=0.5 type=log logbase=2 logstyle=logexpand grid max=64
         labelattrs=(size=8) valueattrs=(size=7)
         label='                                       Relative Risk with 95% CL';
  yaxis display=(nolabel noticks);
run;
```

12.15 LFT Patient Profile

This graph shows LFT patient profiles for the data shown below using the SGPANEL procedure.
- The default panel layout is used with patient as the class variable.
- A band plot is used to display the range of the study days.

patient	alat	biltot	alkph	asat	days	dval	sdays
Patient:5152 White Male Age: 48; Drug: A	0.50000	0.40000	0.40000	0.50000	-25	.	.
Patient:5152 White Male Age: 48; Drug: A	1.36533	0.64286	0.68972	1.08498	0	-0.5	0
Patient:6416 White Male Age: 64; Drug: A	1.50000	1.50000	0.50000	1.00000	-25	.	.
Patient:6416 White Male Age: 64; Drug: A	1.67422	1.63046	0.76092	1.08711	-10	.	.
Patient:6969 White Female Age: 48; Drug: B	0.50000	0.30000	0.70000	0.60000	-25	.	.
Patient:6969 White Female Age: 48; Drug: B	0.71729	0.45636	0.83046	0.79538	0	-0.5	0

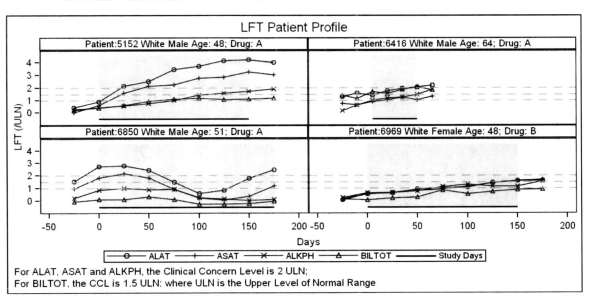

```
title "LFT Patient Profile";
footnote1 j=l "For ALAT, ASAT and ALKPH, the Clinical Concern Level is 2 ULN;";
footnote2 j=l "For BILTOT, the CCL is 1.5 ULN: where ULN is the Upper Level of ..";
proc sgpanel data=sgbook.lft cycleattrs;
  panelby patient / novarname;
  series x=days y=alat / markers lineattrs=(thickness=1px pattern=solid) name='a';
  series x=days y=asat / markers lineattrs=(thickness=1px pattern=solid) name='b';
  series x=days y=alkph / markers lineattrs=(thickness=1px pattern=solid) name='c';
  series x=days y=biltot / markers lineattrs=(thickness=1px pattern=solid) name='d';
  series x=days y=dval / lineattrs=(thickness=2px pattern=solid) name='e';
  band x=sdays lower=dval upper=4.75 / transparency=0.9;
  refline 1 1.5 2 / axis=Y lineattrs=(pattern=dash);
  colaxis min=-50 max= 200;
  rowaxis label="LFT (/ULN)";
  keylegend 'a' 'b' 'c' 'd' 'e';
run;
```

12.16 Panel of LFT Values

This graph shows a classification panel of LFT values by lab test.

- The default panel layout is used with Group as the class variable.
- Each cell contains a scatter plot of sl * ps by drug.
- Three reference lines are added on each axis to display the clinical concern levels.

Drug	Group	PI	SI
Drug A (N=209)	ALAT	0.21429	0.66250
Drug B (N=405)	ALAT	-0.02798	0.67263
Drug A (N=209)	ASAT	0.65714	0.81250
Drug B (N=405)	ASAT	0.44884	1.17656
Drug A (N=209)	BILTOT	0.78857	0.97500
Drug B (N=405)	BILTOT	1.32225	1.08905
Drug A (N=209)	ASAT	0.64286	0.77500
Drug B (N=405)	ASAT	0.58379	0.67525

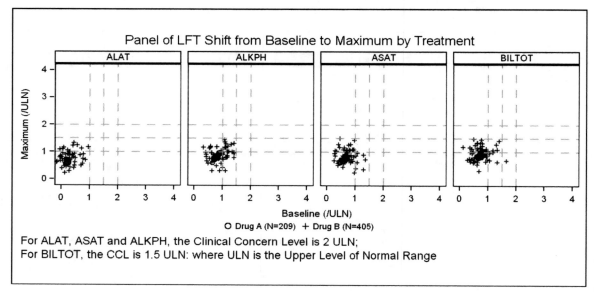

```
title "Panel of LFT Shift from Baseline to Maximum by Treatment";
footnote1 j=l "For ALAT, ASAT and ALKPH, the Clinical Concern Level ...";
footnote2 j=l "For BILTOT, the CCL is 1.5 ULN: where ULN is the Upper...";
proc sgpanel data=sgbook.Panel_LFT pad=(top=40 bottom=40);
  panelby group / layout=panel columns=4 spacing=10 novarname;
  scatter x=pl y=sl / group=drug;
  refline 1 1.5 2 / axis=Y lineattrs=(pattern=dash);
  refline 1 1.5 2 / axis=X lineattrs=(pattern=dash);
  rowaxis integer min=0 max=4;
  colaxis integer min=0 max=4;
  keylegend / title="" noborder;
run;
```

12.17 Distribution of Eye Irritation Using PROC SGPANEL (9.3)

This graph shows the distribution of eye irritation by time and treatment. We have used PROC SGPANEL to simulate the clustered grouped bar chart:

- We use a OnePanel ColumnLattice with class=time.
- Uniscale=Row is set to allow each cell to display only the categories it has.
- The bar labels are displayed using a scatter plot with MarkerChar option.

PARAM	PERCENT	time	chartvar	lcl	ucl	n
EYES ITCHY/GRITTY	40.0000	Week 1	Placebo	25.6864	54.3136	45
EYES ITCHY/GRITTY	60.8696	Week 1	Drug A	49.3541	72.3850	69
EYES ITCHY/GRITTY	52.0548	Week 1	Drug B	40.5947	63.5149	73
EYES ITCHY/GRITTY	62.2222	End Point	Placebo	48.0567	76.3878	45
EYES ITCHY/GRITTY	69.5652	End Point	Drug A	58.7083	80.4221	69
EYES ITCHY/GRITTY	57.5342	End Point	Drug B	46.1954	68.8731	73

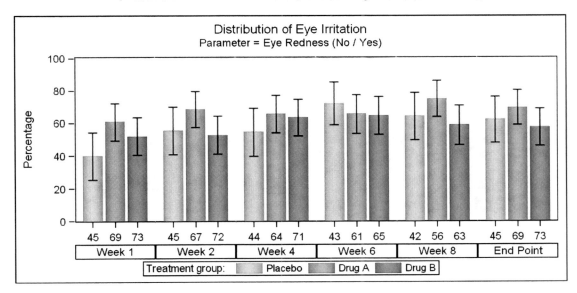

```
proc sgpanel data=sgbook.EyeIrritation;
  format percent 4.1;
  panelby time / layout=columnlattice onepanel novarname
              noborder colheaderpos=bottom uniscale=row;
  vbarparm category=n response=percent / group=chartvar datalabel
         dataskin=pressed;
  scatter x=n y=percent / group=chartvar yerrorupper=ucl markerattrs=(size=0)
         yerrorlower=lcl groupdisplay=cluster;
  colaxis display=(nolabel);
  rowaxis grid max=100;
run;
```

12.18 Distribution of Eye Irritation Using PROC SGPLOT (9.3)

This graph shows the distribution of eye irritation by time and treatment using the SGPLOT procedure. We have used the Y-Y2 split technique to display the statistics at the top:
- We used the VBarParm statement with cluster groups and error bars on the y-axis.
- For stats, use multiple scatter plots of lbl x time with MarkerChar option on the Y2 axis.
- We use OffsetMax=0.25 for Y axis and OffsetMin-0.8 on Y2 axis.

PARAM	PERCENT	time	chartvar	lcl	ucl	n	Nlbl	Lcllbl	UclLbl	PctLbl
EYES ITCHY/GRITTY	40.0	Week 1	Placebo	25.7	54.3	45	N	LCL	UCL	%
EYES ITCHY/GRITTY	60.9	Week 1	Drug A	49.4	72.4	69	N	LCL	UCL	%
EYES ITCHY/GRITTY	52.1	Week 1	Drug B	40.6	63.5	73	N	LCL	UCL	%
EYES ITCHY/GRITTY	62.2	End Point	Placebo	48.1	76.4	45	N	LCL	UCL	%
EYES ITCHY/GRITTY	69.6	End Point	Drug A	58.7	80.4	69	N	LCL	UCL	%
EYES ITCHY/GRITTY	57.5	End Point	Drug B	46.2	68.9	73	N	LCL	UCL	%

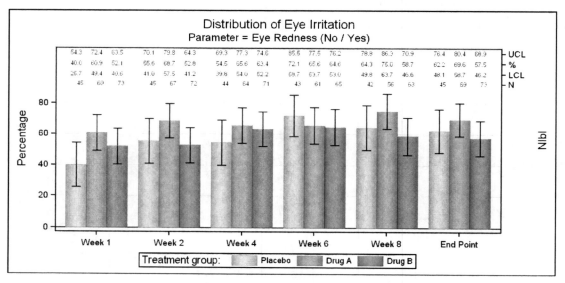

```
title 'Distribution of Eye Irritation';
title2 h=9pt 'Parameter = Eye Redness (No / Yes)';
proc sgplot data=sgbook.EyeIrritationstats;
  vbarparm category=time response=percent / group=chartvar limitupper=ucl
           limitlower=lcl dataskin=pressed;
  scatter x=time y=nlbl / group=chartvar groupdisplay=cluster markerchar=n Y2axis;
  scatter x=time y=lcllbl / group=chartvar groupdisplay=cluster markerchar=lcl Y2axis;
  scatter x=time y=pctlbl / group=chartvar groupdisplay=cluster markerchar=percent
          Y2axis;
  scatter x=time y=ucllbl / group=chartvar groupdisplay=cluster markerchar=ucl Y2axis;
  xaxis display=(nolabel);
  yaxis grid offsetmax=0.25;
  y2axis offsetmin=0.8 offsetmax=0.03; run;
```

12.19 Vital Signs by Time Point Name

This is a graph of some of the vital signs of a patient in a study. The data is extracted from CDISC form and plotted over time by test. In this case we have included only the blood pressure and pulse readings as shown in the table below.
- We use the SGPANEL procedure with Layout=RowLattice with class var=vstestcd.
- We set Uniscale=Column to ensure each row gets its own data range.
- We used a band plot to add the normal ranges for easy evaluation.

vstestcd	VISITDY	VSSTRESN	vstpt	low	high
DIABP	-7	68	AFTER LYING DOWN FOR 5 MINUTES	80	100
DIABP	-7	68	AFTER STANDING FOR 1 MINUTE	80	100
DIABP	-7	62	AFTER STANDING FOR 3 MINUTES	80	100
PULSE	-7	54	AFTER LYING DOWN FOR 5 MINUTES	.	.
PULSE	-7	60	AFTER STANDING FOR 1 MINUTE	.	.
SYSBP	-7	122	AFTER STANDING FOR 1 MINUTE	120	140
SYSBP	-7	134	AFTER STANDING FOR 3 MINUTES	120	140

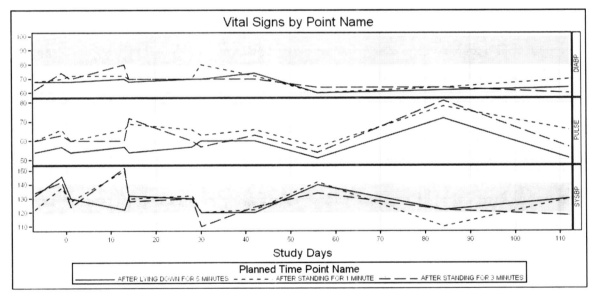

```
title 'Vital Signs by Point Name';
proc sgpanel data=sgbook.vitalsigns;
  where vstestcd <> 'HEIGHT' and vstestcd <> 'TEMP' and vstestcd <> 'WEIGHT';
  panelby vstestcd / layout=rowlattice onepanel novarname uniscale=column;
  band x=visitdy lower=low upper=high / fill nooutline transparency=0.6;
  series x=visitdy y=vsstresn / group=vstpt;
  colaxis grid values=(0 to 120 by 10) valueshint;
  rowaxis grid display=(nolabel);
run;
```

12.20 Concomitant Medications

This graph displays the medications administered to a patient in the study over time and by medication name. The data is shown in the table below.
- We use the SGPLOT procedure and turn off automatic attribute cycling.
- Use a vector plot to draw the duration for each medication.
- We use two scatter plots to label the start and end events for the medications.
- We use a scatter plot with data label to display the medication name.

STARTDATE	STARTDAY	ENDDAY	cmtrt	CMMED	Y	lblday
30MAR13	27	112	BECONASE	BECONASE 1 SPRAY	0.0	27
16FEB13	-48	-1	GINKGO BILOBA	GINKGO BILOBA 1 Tab	0.1	-1
30MAR13	24	112	HYDROCORTISONE	HYDROCORTISONE 1 Unit	0.2	24
16FEB13	-7	112	HYTRIN	HYTRIN 5 Mg	0.3	-7

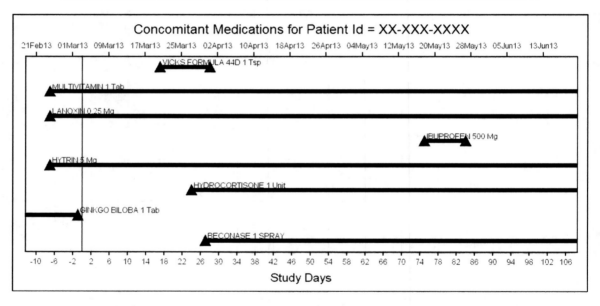

```
title "Concomitant Medications for Patient Id = XX-XXX-XXXX";
proc sgplot data=meds noautolegend nocycleattrs;
  refline 0 / axis=x lineattrs=(thickness=1 color=black);
  vector x=STARTDAY y=y / xorigin=ENDDAY yorigin=y group=cmtrt noarrowheads
      lineattrs=(thickness=5px pattern=solid);
  scatter x=STARTDAY y=y / group=cmtrt markerattrs=(size=13px symbol=trianglefilled);
  scatter x=ENDDAY y=y / group=cmtrt markerattrs=(size=13px symbol=trianglefilled);
  scatter x=STARTDATE y=y /  markerattrs=(size=0) x2axis;
  scatter x=lblday y=y / markerattrs=(size=0) datalabel=cmmed;
  xaxis grid label='Study Days' offsetmin=0.02 offsetmax=0.02
      values=(&minday2 to &maxday by 2);
  x2axis notimesplit display=(nolabel) offsetmin=0.02 offsetmax=0.02
      values=(&mindate2 to &maxdate);
  yaxis display=(nolabel noticks novalues) min=0;
run;
```

12.21 Creating a 2 x 2 Cell Graph Using PROC SGPLOT

Previously, we have used either X-X2 or Y-Y2 split to create two-cell graphs. Here we take the next step and use the X-X2 and Y-Y2 split technique together to create a four-cell (2 x 2) graph using the SGPLOT procedure. Important features of the graph are as follows:

- X and Y axes are restricted to the lower 75% of the space (OffsetMax=0.25).
- X2 and Y2 axes are restricted to the upper 15% of the axis space (OffsetMin=0.85).
- Each graph is drawn using the appropriate combinations of X, Y, X2, and Y2 axes.
- Reference lines are used to demarcate the cells.

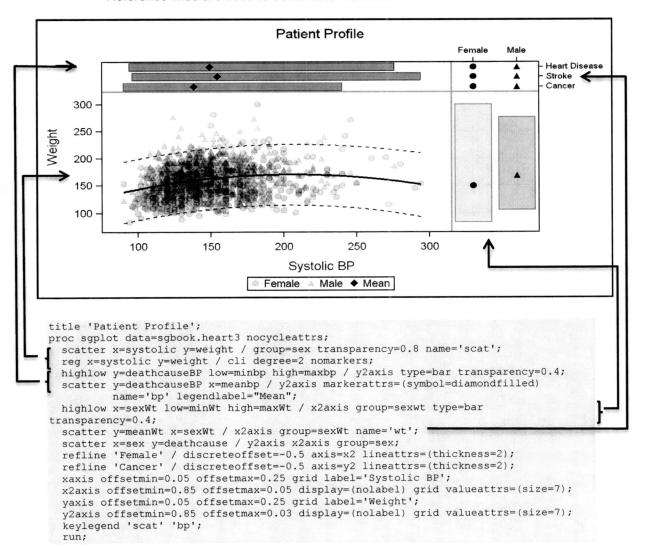

```
title 'Patient Profile';
proc sgplot data=sgbook.heart3 nocycleattrs;
  scatter x=systolic y=weight / group=sex transparency=0.8 name='scat';
  reg x=systolic y=weight / cli degree=2 nomarkers;
  highlow y=deathcauseBP low=minbp high=maxbp / y2axis type=bar transparency=0.4;
  scatter y=deathcauseBP x=meanbp / y2axis markerattrs=(symbol=diamondfilled)
          name='bp' legendlabel="Mean";
  highlow x=sexWt low=minWt high=maxWt / x2axis group=sexwt type=bar
transparency=0.4;
  scatter y=meanWt x=sexWt / x2axis group=sexWt name='wt';
  scatter x=sex y=deathcause / y2axis x2axis group=sex;
  refline 'Female' / discreteoffset=-0.5 axis=x2 lineattrs=(thickness=2);
  refline 'Cancer' / discreteoffset=-0.5 axis=y2 lineattrs=(thickness=2);
  xaxis offsetmin=0.05 offsetmax=0.25 grid label='Systolic BP';
  x2axis offsetmin=0.85 offsetmax=0.05 display=(nolabel) grid valueattrs=(size=7);
  yaxis offsetmin=0.05 offsetmax=0.25 grid label='Weight';
  y2axis offsetmin=0.85 offsetmax=0.03 display=(nolabel) grid valueattrs=(size=7);
  keylegend 'scat' 'bp';
run;
```

Chapter 13

Business Graphs

13.1　Introduction　311
13.2　Stock Price and Volume Chart　312
13.3　Financial Trend and Bond Maturity Graph (9.3)　313
13.4　Danger of High P/E Ratios　314
13.5　Oil Consumption Trend by Country　315
13.6　Product Sales and Target Graph (9.3)　316
13.7　Social Network (9.3)　317

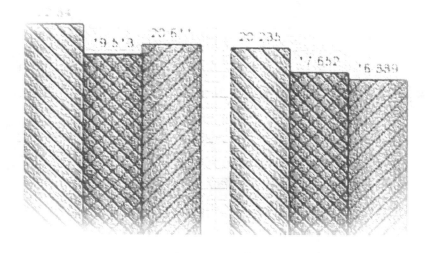

"You can observe a lot just by watching."
~ Yogi Berra

Chapter 13: Business Graphs

13.1 Introduction

The graphs commonly used in the banking, financial, and insurance industries have different requirements than graphs used for the statistical analysis of data. These graphs are more "creative" in nature. While the principles of effective graphics continue to be applicable for these graphs, they also generally need more visual aesthetics.

The graphs shown in this book have been customized for gray-scale rendering. However, color graphs are particularly useful for business graphics, and the visual appearance of these graphs can be further enhanced by the use of color styles.

The techniques described in the earlier chapters of this book can be leveraged in creative ways to create such graphs. Here we have used the axis-splitting technique to display multiple graphs in one cell, like the graph for bond yields. We have used combinations of HighLow and VBar to do the "Product Sales and targets" graph, and we used a bubble plot to do the "Social Graph".

In this chapter, we will examine how you can create some common graphs used in this industry. Some examples are shown in Figure 13.1. Detailed code for creating such graphs is shown in the following pages.

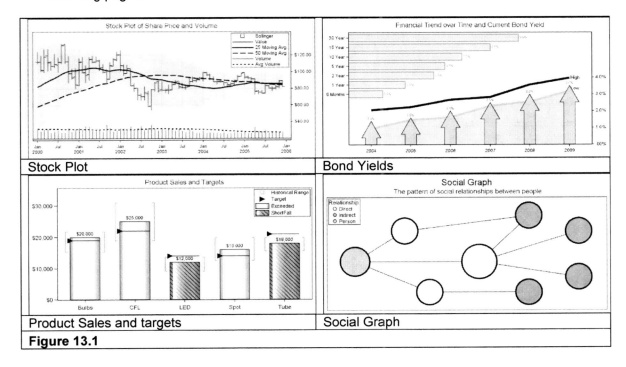

Figure 13.1

13.2 Stock Price and Volume Chart

This is a common stock price chart from the data shown below with the following features:

- Y-Y2 split technique is used to draw price data above the volume data.
- Price data is an overlay of the Bollinger band, 25 and 50 event moving average, and price.
- Volume data includes volume and moving average.
- The key legend is placed inside in the empty upper right corner with all the details.

Stock	Date	Open	High	Low	Close	Volume	avg25	avg50	upper	lower	vavg
IBM	01AUG88	$126.00	$126.87	$110.37	$112	5,256,886	$23	.	$29	$16	7560307.56
IBM	01SEP88	$111.12	$116.50	$109.50	$115	5,433,352	$22	.	$29	$16	7552677.08
IBM	03OCT88	$115.12	$124.87	$112.75	$123	5,790,742	$22	.	$29	$16	7462149.84

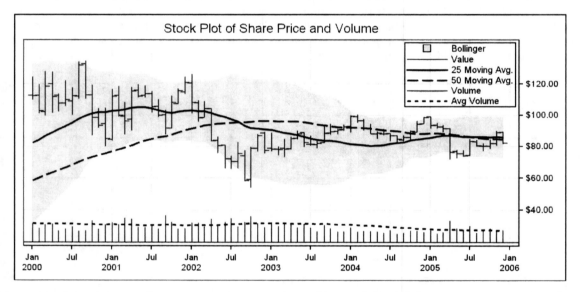

```
title 'Stock Plot of Share Price and Volume';
proc sgplot data=sgbook.Stock_IBM_Stats;
  where date > '01Jan2000'd;
  band x=date upper=upper lower=lower / y2axis transparency=0.5 legendlabel='Boll…';
  highlow x=date high=high low=low / y2axis open=open close=close
          lineattrs=(color=black thickness=1) legendlabel='Value';
  series x=date y=avg25 / y2axis lineattrs=graphdata1 legendlabel='25 Moving Avg.';
  series x=date y=avg50 / y2axis lineattrs=graphdata2 legendlabel='50 Moving Avg.';
  needle x=date y=volume / lineattrs=graphdata3(pattern=solid);
  series x=date y=vavg / lineattrs=graphdata4(opts) legendlabel='Avg Volume';
  keylegend / location=inside position=topright across=1 valueattrs=(size=8);
  yaxis max=100000000 display=none;
  y2axis display=(nolabel) grid min=20 values=(40 to 160 by 20) valueshint;
  xaxis display=(nolabel) grid;
run;
```

13.3 Financial Trend and Bond Maturity Graph (9.3)

This graph shows the financial trend over time and bond yields by maturity. The features of the graphs are:

- X-X2 and Y-Y2 split technique is used to display the financial trend and yield in the same graph.
- The trend is drawn in the X-Y2 space using two series, one band and one HighLow plot.
- The bond yields are drawn in the X2-Y space using a HighLow plot.

Maturity	Rate	Year	Value	High	Low	Zero
6 Months	.60%	.	.	.	.	0
1 Year	1.0%	.	.	.	.	0
2 Year	1.5%	.	.	.	.	0
5 Year	1.7%	.	.	.	.	0
10 Year	2.0%	.	.	.	.	0
15 Year	2.5%	.	.	.	.	0
30 Year	3.0%	.	.	.	.	0
.	.	2004	1.4%	2.0%	.94%	0
.	.	2005	1.6%	2.2%	1.5%	0
.	.	2006	2.0%	2.7%	1.6%	0
.	.	2007	2.5%	2.8%	2.3%	0
.	.	2008	3.0%	3.6%	2.6%	0
.	.	2009	3.5%	4.0%	3.2%	0

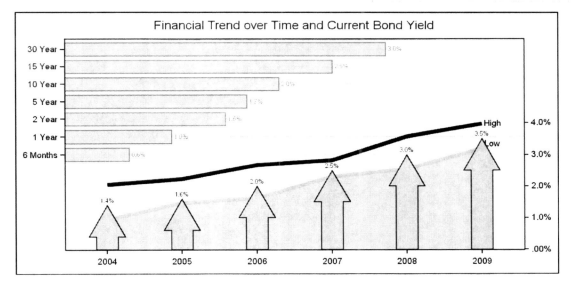

```
title 'Financial Trend over Time and Current Bond Yield';
proc sgplot data=sgbook.Trend noautolegend;
  band    x=year lower=0 upper=low / transparency=0.4 y2axis;
  series  x=year y=high / lineattrs=(thickness=5) y2axis curvelabel='High';
  series  x=year y=low / lineattrs=(thickness=5 color=cxc0c0c0)
          y2axis curvelabel='Low';
  highlow x=year high=value low=zero / highcap=filledarrow y2axis
          type=bar barwidth=0.3 transparency=0.0 highlabel=value;
  highlow y=maturity low=zero high=rate / type=bar barwidth=0.8 x2axis
          highlabel=rate transparency=0.3 ;
  xaxis type=discrete display=(nolabel);
  x2axis offsetmin=0 offsetmax=0.3 display=(nolabel noticks novalues);
  y2axis offsetmin=0 offsetmax=0.4 display=(nolabel);
  yaxis offsetmin=0.45 offsetmax=0.05 type=discrete display=(nolabel) grid;
run;
```

13.4 Danger of High P/E Ratios

This graph displays the price-earnings (PE) ratios for stocks over the last 125 years. The features of the graph are as follows:

pe10	Year	re	pe	re10	label
18.2148	1881	8.7483	12.6327	7.5378	.
15.6153	1882	8.7123	13.4545	7.7364	.
15.2162	1883	8.7730	13.5116	7.8609	.
14.4526	1884	7.5805	12.9500	7.8101	.
13.2753	1885	6.8382	13.6774	7.8221	.
26.7007	1929	19.0579	18.0145	11.9516	1929

- A band plot is used to draw a filled graph of the PE over time.
- A second band plot is drawn to show the PE=20 zone.
- A series plot is drawn to display the actual PE values over time.
- The PE values are displayed using the "Label" variable. This column contains non-missing values only for the ones we want to include in the graph.

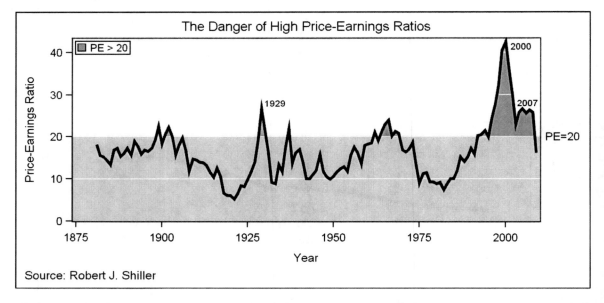

```
title "The Danger of High Price-Earnings Ratios";
footnote j=l "Source: Robert J. Shiller";
proc sgplot data=sgbook.PE_DATA noautolegend;
  band x=year upper=pe10 lower=0 / fillattrs=graphdata2
       name='high' legendlabel='PE > 20';
  band x=year  upper=20 lower=0;
  refline 20 / label="PE=20";
  refline 10 20 30 40 / lineattrs=GraphGridLines;
  keylegend 'high' / location=inside position=topleft;
  series x=year y=pe10 / lineattrs=GraphData1(thickness=3) datalabel=label;
  yaxis offsetmin=0;
run;
```

13.5 Oil Consumption Trend by Country

This graph displays the oil consumption trend over 30 years by country as a percentage of 2005 consumption. The data is shown in the table.

Year	Country	value	total	subtotal	subtotalpct	valuepct	valuelabel
1965	USA	11522	29524	11992	40%	38%	
1965	China	217	29524	11992	40%	1%	
1965	India	253	29524	11992	40%	1%	
2005	USA	20655	77099	30128	100%	69%	69%
2005	China	6988	77099	30128	100%	23%	23%
2005	India	2485	77099	30128	100%	8%	8%

- The SGPANEL procedure is used with PANELBY = country.
- The empty bar chart plots the combined trend for all three countries.
- The filled bar chart plots the consumption of the individual country as a percentage of the total.
- A bar chart is used to display the bar label for just the most recent year.

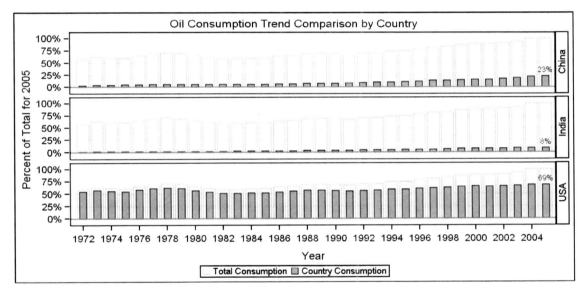

```
title 'Oil Consumption Trend Comparison by Country';
proc sgpanel data=sgbook.oil nocycleattrs;
  where country NE 'Japan' and year > 1971;
  panelby country / columns=1 onepanel layout=rowlattice novarname spacing=6;
  vbar year / response= subtotalpct transparency=0.8
      legendlabel='Total Consumption' name="a" fillattrs=graphdata5 ;
  vbar year / response= valuepct transparency=0.2 fillattrs=graphdata2
      legendlabel='Country Consumption' name="b" barwidth=0.5 ;
  vbar year / response= valuelabel datalabel fillattrs=graphdata2;
  rowaxis grid label='Percent of Total for 2005' values=(0 to 1 by 0.25);
  colaxis fitpolicy=thin;
  keylegend "a" "b";
run;
```

13.6 Product Sales and Target Graph (9.3)

This graph displays sales by product, with target and the historical low and high range. The data is shown on the right.

Status	Sales	Target	High	Low	Product	High2	Low2
Exceeded	$20,000	$19,000	$21,768	$15,256	Bulbs	23945.27	13730.65
Exceeded	$25,000	$22,000	$30,991	$18,882	CFL	34089.85	16993.41
ShortFall	$12,000	$14,000	$13,624	$9,174	LED	14986.82	8256.79
Exceeded	$16,000	$14,000	$19,844	$12,535	Spot	21828.51	11281.64
ShortFall	$18,000	$21,000	$19,882	$15,484	Tube	21870.44	13935.41

- A grouped VBARPARM statement is used to display the sales data.
- Two HIGHLOW plots are used to draw the historical range for each product.
- A SCATTER plot is used to draw the target arrow.
- Another HIGHLOW bar is used to draw the target line.
- All items are displayed in an inside legend.
- Note: Some hard-coded colors that work for this style may need adjusting.

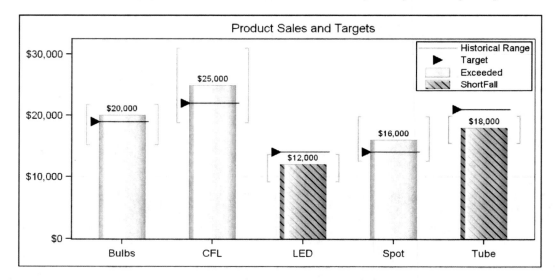

```
title 'Product Sales and Targets';
proc sgplot data=sgbook.target;
  highlow x=product high=high low=low / type=bar barwidth=0.8 transparency=0.5
        name='range' legendlabel='Historical Range' nofill;
  highlow x=product high=high2 low=low2 / type=bar barwidth=0.7
        lineattrs=(color=white thickness=2) nofill;
  vbarparm category=product response=sales / group=status datalabel
        dataskin=gloss barwidth=0.6 name='bar';
  highlow x=product high=target low=target / type=bar barwidth=0.6
        nofill lineattrs=(color=black);
  scatter x=product y=target / markerattrs=(symbol=trianglerightfilled size=12)
        discreteoffset=-0.3 name='target' legendlabel='Target';
  keylegend 'range' 'target' 'bar' / location=inside position=topright across=1;
  yaxis display=(nolabel) offsetmin=0 grid;
  xaxis display=(nolabel);
run;
```

13.7 Social Network (9.3)

This graph displays social relationships between persons in a social network.

- A BUBBLE statement is used to draw the nodes, with a size variable.
- A VECTOR statement is used to draw the links. The arrowheads are turned off.
- A legend is used to show the type of relationship.

Node	x	y	size	Group	Link	x1	y1	x2	y2
A	-1.5	2	1.0	Direct	A	-2.5	0	0.0	0
B	1.0	3	1.0	Indirect	B	-2.5	0	-1.5	2
C	2.0	2	1.0	Indirect	C	-1.5	2	1.0	3
D	2.0	-1	1.0	Indirect	D	0.0	0	1.0	3
E	1.0	-2	1.0	Indirect	F	0.0	0	2.0	2
F	-1.0	-2	1.0	Direct	G	0.0	0	2.0	-1
G	-2.5	0	1.1	Person	H	-2.5	0	-1.0	-2
H	0.0	0	1.3	Direct	I	-1.0	-2	1.0	-2

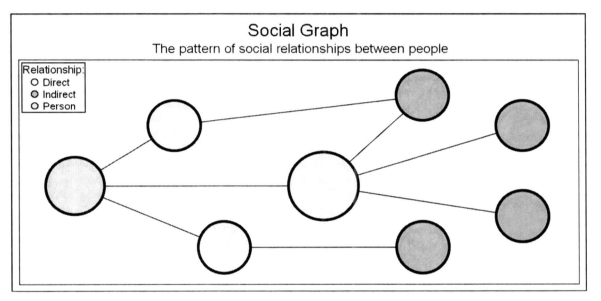

```
title h=14pt 'Social Graph';
title2 h=11pt ' The pattern of social relationships between people';
proc sgplot data=sgbook.social;
   vector x=x2 y=y2 / xorigin=x1 yorigin=y1 lineattrs=(thickness=1)
           noarrowheads;
   bubble x=x y=y size=size / group=group bradiusmin=30 bradiusmax=40
           fill outline lineattrs=(thickness=3) name='b';
   keylegend 'b' / title='Relationship:' location=inside position=topleft
           across=1;
   yaxis display=none;
   xaxis display=none offsetmin=0.1 offsetmax=0.1;
   run;
```

Chapter 14

Styles

14.1 Introduction 321
14.2 Using Styles 321
14.3 Style Elements 321
14.4 Using Style Elements 324
14.5 Style Element Usage Precedence 324

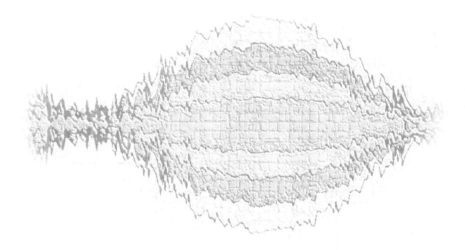

"The theory of probabilities is at bottom nothing but common sense reduced to calculus."
~Laplace

Chapter 14: Styles

A brief introduction to usage of styles with ODS Graphics was included in Section 2.5. Let us review that information and expand on the topic of styles and their usage.

14.1 Introduction

Graphs created by the SG procedures derive their visual look and feel from the active style for the output destination. Each ODS destination has a default associated style. Styles supplied by SAS are designed to ensure effectiveness and an aesthetically pleasing appearance by default. Default styles for common destinations are shown in Figure 14.1.

You can create your own style to suit the look you prefer or to present a consistent corporate look by using the TEMPLATE procedure. The description of PROC TEMPLATE is beyond the scope of this book. However, since styles are intimately related to the topic of graphs, we will cover the related relevant topics here.

ODS Destination	Style Name
LISTING	Listing
HTML	HtmlBlue
RTF	RTF
PDF	PRINTER

Figure 14.1

For a discussion of styles and the TEMPLATE procedure, see the SAS documentation.

14.2 Using Styles

To change the overall visual appearance of the graphs, you can assign a different style by using the STYLE option on the ODS destination statement.

>   ods *destination* style=*style-name*;

When you do this, all subsequent output sent to the destination will use this assigned style until the destination is closed or the style is changed.

14.3 Style Elements

A style is a collection of named elements. Each element is a bundle of named visual attributes such as color, font, marker symbols, and so on. Output tables derive visual attributes such as fonts, background colors for the headers, size, color of titles, etc. from the style.

The default visual attributes of various graph elements are also derived from the style. Plot colors, marker symbols, line thickness, axis label fonts, etc. are derived from specific named elements of the style. The association between the element of the graph and the style element is well defined and described in detail in the product documentation. Some common style element names and their associated graph elements are shown in Figures 14.2 and 14.3.

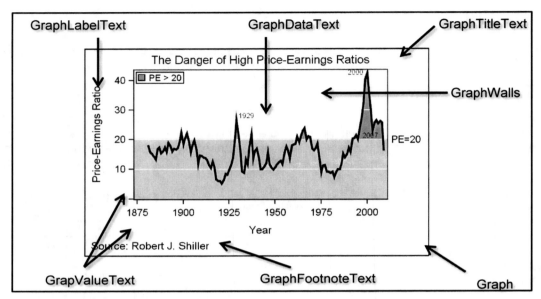

Figure 14.2: Graph and Text Elements

Figure 14.2 above lists some style elements that are commonly used for various elements in the graph. These elements are used to determine the visual attributes for the graph background, walls, and the various text statements such as titles, footnotes, etc. The elements are listed below. This graph was created with SAS 9.3 PROC SGPLOT.

Element Name	Usage
Graph	Graph background fill and border
GraphWalls	Wall background fill and border
GraphTitleText	Title font, size, color, etc.
GraphFootnoteText	Footnote font, size, color, etc.
GraphLabelText	Axis Label font, size, color, etc. Also Legend Title
GraphValueText	Axis tick value and Legend entry font, size, color, etc.
GraphDataText	Data labels font, size, color, etc.

If you want to create a new style that has smaller fonts for the tick values, you can customize the GraphValueText element appropriately. Then, all the graphs using this style will display the new font you have selected.

A new style can be created using the TEMPLATE procedure. It is generally good practice to derive a new style from a parent style. In this way, you only need to customize the elements that you need to change. For details on how to create a new style, see the SAS documentation for the TEMPLATE procedure.

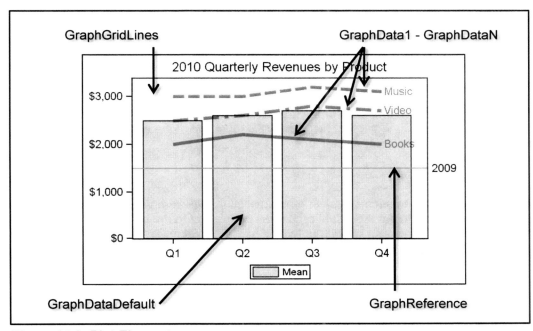

Figure 14.3: Plot Elements

Figure 14.3 above shows some style elements that are commonly used for various plot elements in the graph. These elements are used to determine the visual attributes for the parts of the bar chart or series plot.

The GraphDataDefault element is used to display plots that do not have a GROUP option. In Figure 14.3, the bar chart does not have a GROUP option set, so all the bars for this plot use the GraphDataDefault style element.

Plot statements that use a GROUP option utilize the visual attributes from GraphData1 to GraphData12 style elements. Most styles supplied by SAS have 12 data elements. However, there is no limit, and you can define more or fewer elements as you please. For a full list, see the SAS documentation.

Element Name	Usage
GraphDataDefault	Non-Grouped data elements
GraphData1 - N	Grouped data elements – or when using CycleAttrs
GraphFit	Fit lines
GraphFit2	A second element for Fit lines, not used by default
GraphReference	Reference lines
GraphGrid	Grid lines
GraphConfidence	Confidence plots

14.4 Using Style Elements

If you customize and use your own style, this new style is used for all procedure output to the ODS destination, until the destination is closed, or the style is changed again. Often you do not want to change the style setting of a destination, but only want to make a specific change for one graph.

For example, the default font size for the x-axis tick values may be just a little too big to fit the space available. Shrinking the font just a bit will do just fine. One way to do this would be to change the size of the currently used font:

    xaxis valueattrs=(size=7);

You will likely have seen such syntax in many places in the examples. This works fine, but in other cases you want to set the visual properties of one plot to the same as another item in the graph. You can use the Data Label font for the Axis Tick Values as follows:

    xaxis valueattrs=GraphDataText;

This technique is especially useful when using color output. If you have two separate measures plotted on the same graph, one to the y-axis and one to the Y2 axis, it can be useful to set the color of the axis tick values to the same color as the plot line. You can do that as follows:

    yaxis   valueattrs=GraphData1;
    y2axis valueattrs=GraphData2;

14.5 Style Element Usage Precedence

When you use a color style, hard-coded color values are not recommended. This is because the value specified on the plot option wins over the visual attributes derived from the ODS style. You may set the color of a line to "Black" because the wall color of the current style is white. But if the same graph is run with a style that has a dark wall color, this "Black" color will blend into the wall color, resulting in a bad color choice for the graph.

The precedence for usage of visual attributes is as follows:
1. First the procedure derives the visual attributes from active Style.
2. This is replaced by attributes from Discrete Attribute Map (if applicable).
3. This is replaced by visual attributes specified on the statement option.

For the reason above, and to ensure your graph works well for all use cases, it is recommended you always use "style relative" settings. If you do not like a color used in the graph, try one of the other colors defined in the style. Hard-coded colors are best used for one-off graphs that are not likely to be used with different style settings (see Figure 13.6).

Chapter 15

ODS Destination and ODS Graphics Options

15.1 Introduction   327
15.2 ODS Destination Options   327
15.3 ODS Graphics Options   328

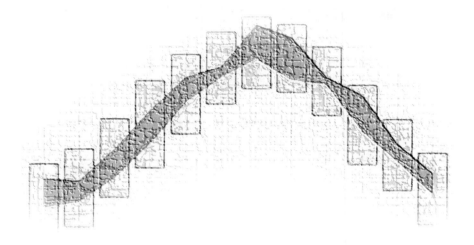

"Not everything that can be counted counts,
and not everything that counts can be counted."
~ Albert Einstein

Chapter 15: ODS Destination and ODS Graphics Options

15.1 Introduction

All graph output from SG procedures is sent to the open ODS destinations. If multiple destinations are open, output is sent to each destination using the various options that are in effect for each destination.

Global options that apply to the graphs are specified on the ODS GRAPHICS statement. In Chapter 2, we discussed the use of the ODS GRAPHICS statement to switch on the creation of automatic graphs from the SAS procedures. With SAS 9.3, ODS Graphic is ON by default for all procedures including the SG Procedures. Global options that control certain aspects of the graphs can be provided on this statement. These options are discussed below.

15.2 ODS Destination Options

The options relevant to graphs are discussed in Figure 15.1 below. For complete details on all options for each ODS destination, see the SAS documentation for ODS Language Statements.

Option	Values	Description
DPI	Value	DPI for generated document
GPATH	Location	The location for all graphics output for this destination.
IMAGE_DPI	Value	DPI for generated image.
PATH	Location	The location for the output file for this destination.
SGE	OFF | ON	Enable / disable creation of editable graph output
STYLE	Style name	The ODS Style to be used for this destination.

Figure 15.1

SGE option: This option is used to enable or disable the creation of editable graphical output in the format used by the ODS Graphics Editor. For SAS 9.2, only the LISTING destination supports this option. For SAS 9.3 this option is available on all ODS destinations.

By default, the SGE option is OFF. When SGE=ON, an editable graph is created along with the image file. This output is listed in the Results window. You can open this file to launch the ODS Graphics Editor to make non-persistent customizations to the ODS Graphics output.

DPI and Image_DPI: Graphs are rendered at default DPI for the output destination. For some destinations, the DPI applies only to the graph image. The size of the graph is based on the active DPI. If the size is specified in pixels (the default is in pixels), then the baseline for scaling is the active DPI for the destination.

15.3 ODS Graphics Options

Certain attributes of the ODS graphics output are controlled by options on the ODS GRAPHICS statement. These options stay in effect until reset or the ODS Graphics system is turned off. The syntax for the statement is as follows:

```
ods graphics <on | off> </ options;
```

The options are listed in Figure 15.2 below.

Option	Values	Description
ANTIALIAS	ON | OFF	Enable or disable anti-aliasing of lines and markers
ANTIALIASMAX	Value	Maximum number of observations for anti-aliasing
BORDER	ON | OFF	Display or hide the graph border
DISCRETEMAX	Value	Maximum number of discrete values
GROUPMAX	Value	Maximum number of group values
HEIGHT	Value	Height of the graph in px, in, cm, etc.
OUTPUTFMT	PNG, GIF, etc.	Any one of the supported formats for the destination
IMAGEMAP	ON | OFF	Enable or disable creation of image map for HTML
IMAGENAME	string	Name for the output file
LABELMAX	Value	Maximum number of labels displayed
MAXLEGENDAREA	Value	Maximum % of the height occupied by legends
PANELCELLMAX	Value	Maximum number of cells in a panel
RESET		Reset all (or individual) options to default
SCALE	ON | OFF	Enable or disable scaling of graph elements
TIPMAX	Value	Maximum number of tips displayed
WIDTH	Value	Width of the graph in px, in, cm, etc.

Figure 15.2

For complete details on all options for each ODS destination see the SAS documentation for ODS Language Statements.

Anti-aliasing: This is a technique used to improve the rendering of various elements in the graph such as the markers, lines, text, and so on. Anti-aliasing improves the visual appearance of the elements but consumes more computing resources. Text elements in the graph are always anti-aliased for high quality. Anti-aliasing of the line and marker elements can be controlled by these options. When there are too many observations in the data, anti-aliasing becomes less effective and more costly. By default, anti-aliasing is turned off for markers and lines when the number of observations exceeds 600. The ANTIALIAS option can be used to enable or disable anti-aliasing. The ANTIALIASMAX option can be used to change the level when anti-aliasing is turned off.

DISCRETEMAX, GROUPMAX: If the number of discrete values in a graph, such as number of categories on the axis or the number of group values becomes large, the procedure can slow down or even run out of memory. This can happen if the user inadvertently selects a variable that has a large number of discrete values for a discrete role, such as zip code, etc. In normal usage, variables used for such roles have < 1000 unique values. In this case, the procedure will stop processing and log a message. If you really need to run such a use case, you can set these settings to a larger value to run your use case.

OUTPUTFMT: The default output format is determined by the ODS destination. For many destinations like LISTING, HTML, etc., the graph is created in industry-standard image formats. You can specify the format you want by using this option. The value you provide will only be honored if the output destination supports the format.

IMAGEMAP: For the HTML destination, tool tips can be displayed for the data values in the graph. By default, this setting is OFF since it consumes computing resources. You can enable creation of the image map for display of tool tips for HTML destinations by using this option.

IMAGENAME: The default name for the graph output file is "sgplot.xxx" or "sgpanel.xxx" based on the procedure used. File names are automatically appended with an incremented counter to avoid overwriting of the files from multiple procedure executions. You can provide your own preferred name for the filename prefix.

LABELMAX: Often you can request the display of data labels with a scatter plot. The procedures will attempt to place the data labels close to the markers, with minimal collision between labels. At some point, there are too many labels in the graph, and the graph becomes ineffective. By default, the display of labels is disabled when the number of labels exceeds 200. You can use this limit to control this feature.

MAXLEGENDAREA: By default, the legends for the graph are dropped if the area occupied by them exceeds 20% of the total area of the graph. This is to prevent cases where the legend is so big that there is not much space left for the display of the data itself. You can control this limit by using this option.

PANELCELLMAX: The SGPANEL procedure subdivides the graph into cells based on the number crossings of the classification variables. By default, if the number of cells exceeds 10,000, the graph is not created and a message is logged. You can control this limit by using this option.

RESET: You can reset all the options to the default settings by using this option. You can also use `reset=option-name` to reset one or more specific options. You can use `reset=index` to reset just the counter that is used to postfix an index to the output file name.

SCALE: The output graph size is determined by the procedure. See the WIDTH or HEIGHT option. If you request a different height of the graph, the internal details of the graph are scaled using a nonlinear algorithm designed to render graphs with reasonable font sizes. In such a case, the actual font size may be different from what was specified for the text in the graph. You can enable or disable this scaling for height by using this option.

TIPMAX: Tooltips can be displayed for the data values for the HTML destination. By default, if the number of observations exceeds 500, tooltips are disabled. You can control this limit by using this option.

WIDTH and HEIGHT: The default width and height for the graph is determined by the procedure, and is often 640px x 480px. In case of paneled graphs, the size may vary based on the number of cells. You can control the size of the graph using these options.

Chapter 16

Tips for Graph Output

16.1 Introduction 333
16.2 Creating Small Graphs for Use in Documents 333
16.3 Creating Large Graphs for Use in Presentations 334
16.4 Combining Graph Size and DPI 336
16.5 Impact of Graph Size on System Resources 336

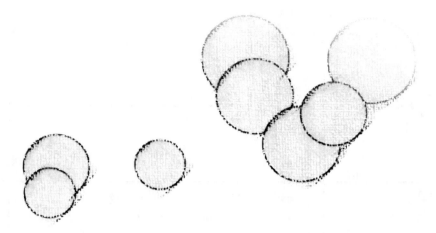

"Each problem that I solved became a rule which served afterwards to solve other problems."
~ Rene Descartes

Chapter 16: Tips for Graph Output

16.1 Introduction

After completing the data analysis, the analyst needs to share the results of the analysis, along with graphs, with various consumers. This means the graphs will be inserted into Web pages, publications, documents and overhead presentations. It is critical that these graphs look their best and deliver the information effectively in the final form.

For the LISTING destination, the default graph size is 640 pixels x 480 pixels at 96 DPI. These default settings work well for viewing the graphs on the computer screen. For inclusion of such graphs in various documents, some customization is useful.

16.2 Creating Small Graphs for Use in Documents

Often you will need to insert a graph into a document where the graph must fit inside a small inset or in one-column of a two-column layout. Figures 16.1 and 16.2 show two graphs, both with a width of 2.875 inches. For Figure 16.1, we have inserted the default output from the SGPLOT procedure into the available region. This default output is rendered as a 640px x 480px image. The publication software linearly scales the image to fit the box, thus shrinking all the graph elements, including the fonts, making the font size less than optimal.

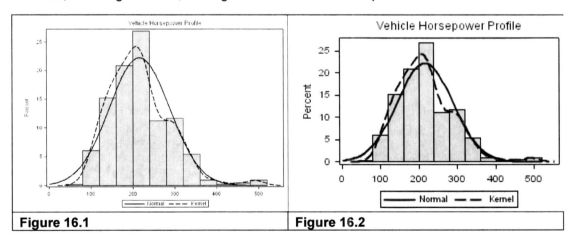

| **Figure 16.1** | **Figure 16.2** |

For the graph for Figure 16.2, we need to fit the graph inside a 2.875-inch-wide box. So, we have created the graph of the correct size by using the ODS Graphics WIDTH option. When this graph is inserted into the document, no image scaling is required.

The graph in Figure 16.2 has fonts that are easy to read. The line thickness is also bolder. If the graph had markers, they would be bigger. Here the graph is doing the scaling, and it uses a font-friendly way to scale these items, resulting in a more readable graph.

Design and Render Size: Each graph has a default "design" size. For most cases, this is 640px x 480px. This size cannot be changed directly in SG procedures, but it can be changed if you are using GTL. A different "render" size can be specified by using the WIDTH and HEIGHT options on the ODS GRAPHICS statement.

When the render size is different from the design size, the graph is scaled to fit the new size. All individual elements inside the graph are scaled in a sub-linear fashion. The scale is:

$$\text{Scale-Factor} = (\text{Render-Height} / \text{Design-Height})^{0.25}$$

This is done specifically to slow the shrinking or growth of the font sizes. Therefore, as the graph height shrinks, the font size shrinks too, but at a slower rate. Similarly, as the graph height grows, the font size grows too, but at a slower rate.

For our use case shown above, Design-Height = 480px. The Render-Width is 2.875. If we retain the aspect ratio, the render width is 2.875 *.0.75 = 2.16in at 96 DPI, resulting in a graph width of 207 pixels. The Linear scale for the image size is 207 / 480 = 0.43. However, the Scale-Factor shown above is 0.43 ** 0.25 = 0.81.

In Figure 16.1, the scale factor for the image and the fonts is 0.43. Therefore, a 10-point font effectively shows as a 4-point font, which is quite small. For the graph in Figure 16.2, the scale factor for image is 0.43, but the scale factor for the fonts is 0.81. So, a 10-point font is drawn at 8 point, which is still quite readable.

Recommendation for Creating Small Graphs: Always render your graph at a size close to its final usage. Using the default size graph in a document will be fine if the graph fills a width of 6 to 7 inches. But if the graph is to fit a smaller size, do not use the default size graph.

Caveat: This does mean that the final rendered font sizes may differ from those specified in the code or the style. If it is important that the font size be exactly as specified, then you can use the NOSCALE option on the ODS GRAPHICS statement. Even if the graph render height is different from the default design height, the fonts, lines, and markers will not scale.

16.3 Creating Large Graphs for Use in Presentations

The dimensions of the output image are also controlled by the DPI or IMAGE_DPI for the destination. Each destination has its own default DPI used by the document or just the image. Some default values are shown in Figure 16.3.

For LISTING and HTML destinations, the graph image is created as a separate image file. In this case, the default image format is PNG, and the DPI information is included in the file.

ODS Destination	Default DPI
LISTING	96
HTML	96
RTF	150
PDF	150
Figure 16.3	

If you want to create graph output that can be embedded in another document or presentation, it is convenient to create a separate file using the LISTING destination. With SAS 9.2, all graph output is in image format. Even for PDF and RTF output, the graph is essentially an image embedded into the document.

With SAS 9.3, support for vector graphics output has been added. You can now generate EMF, PDF, PS, SVG, and PCL output, depending on the ODS destination you use. In some special cases, the graphs use data skins or have gradient legends. In such cases, the output is in image formats. This is also the case when using transparency for EMF and PS.

Graph size is controlled by setting the WIDTH and HEIGHT options on the ODS GRAPHICS statement and DPI, as seen in section 16.2. When creating smaller graphs for embedding into documents, it is useful to change the graph size. It is also useful to increase the graph size when creating a graph that is more detailed in nature and needs more space for rendering all the details.

Recommendations for Creating Large Graphs: When presenting a graph to a large audience using a projection system, it is necessary that the details of the graphs are visible from a distance. The same graph that looks nice on the screen from two feet away should be viewable on the large screen from 50 to 100 feet away. To achieve this, we need to "scale up" the graph so that all aspects of the graph are made bolder and bigger linearly.

Vector format output is inherently scalable and ideal for use in some cases. So you can use the default output in EMF or PDF formats for such use cases. You can enlarge the graph on the computer screen to fill the window, and these vector-based graphs will scale up nicely.

Image-based output created at a low (or default) resolution or DPI does not scale up very well. When scaled, you will see "jaggies" along the edges of the marker, lines and fonts. To include such graphs in a presentation using a projection system to a large audience, you should increase the DPI of the graph. This will make the graph larger and also make all the elements of the graph (for example, markers, line thickness, and fonts) linearly bigger. Therefore, this graph, when viewed from afar, will look readable.

Creating a graph with a size of 6.4in x 4.8in at 200 DPI will generate an image that is 1280px x 960px. The PNG file will include the DPI information in the file; therefore, when you include this graph in a document, it will occupy a region 6.4in x 4.8in by default. However, if you view this file in some image viewers or Internet Explorer, you may see the image in the full 1280 x 960 size because these viewers do not honor the DPI setting in the PNG file.

Recommendations for Graphs in Web Pages: Some Web browsers do not honor the DPI setting in the PNG file. A graph intended to be 6in x 4in (200 DPI) may be shown as 12in x 8in in the browser. Other browsers do scale, but poorly. So, it is best to use the ODS HTML destination, or create graphs at 96 DPI for usage in Web pages.

16.4 Combining Graph Size and DPI

As discussed in sections 16.2 and 16.3, scaling of the graph for height and DPI are done differently. Each is useful for different use cases. You can also mix the two judiciously to achieve the custom look you need. High DPI is also useful for documents since they are usually printed at a higher resolution (300 to 600).

In this book, we have used a size of 3in x 2in at 300 DPI for most of the graphs. Specifying the size shrinks the fonts a little. Using high DPI gives us graphs that will scale well when printed on a high-resolution printer. For graphs that occupy a larger area, like those in Chapters 12 and 13, we have used a larger size of 6in x 3in at 200 DPI.

16.5 Impact of Graph Size on System Resources

When creating graphs as image output, higher system resources are consumed to create larger graphs with high DPI.

- More system memory is needed to create the graph. The rendering system is Java based. SAS system settings for Java specify the amount of memory that can be used by Java. Using higher size and DPI requires a higher amount of memory. If the graph cannot be rendered, a warning is logged. If you need the higher resolution graph, you can change the memory settings using the JREOPTIONS option specified in the SAS system options. See the SAS documentation for details.

    ```
    -jreoptions '(-Xmx512m)'
    ```

- If sufficient memory resources are available, the graph will be rendered. However, it will likely consume more system and CPU time to render the graph.

- If the graph is successfully rendered, the output file size will be bigger.

Index

A

accuracy of magnitude perception 6
ACROSS= option, KEYLEGEND statement (SGPLOT) 231
adverse event timelines 291–292, 300
ALPHA= option
    CLM option 140
    DOT statement (SGPLOT) 215
    ELLIPSE statement (SGPLOT) 153, 155
    HBAR statement (SGPLOT) 194
    HLINE statement (SGPLOT) 208
    LOESS statement (SGPLOT) 140
    PBSPLINE statement (SGPLOT) 146–147
    REG statement (SGPLOT) 132
    VBAR statement (SGPLOT) 186
    VLINE statement (SGPLOT) 201
annotations
    about 235–237
    adding company logos 243
    background images in graphs 244
    bubble legends 248
    creating axis-aligned statistics table 240
    creating axis tick values with Unicode 238
    creating forest plots with 241–242
    creating multi-axis tick values 239
    images as curve labels 245–246
    making polyline figures 249–250
    QTc change graph with annotated at risk values 296
    using arrows to emphasize data 247
ANTIALIAS option, ODS GRAPHICS statement 328
ANTIALIASMAX= option, ODS GRAPHICS statement 44, 79, 328
ARROW function 236
ARROWHEADSHAPE= option, NEEDLE statement (SGPLOT) 99
arrows, emphasizing data with 247
attribute maps 235, 251–253
automatic graphs from procedures 10–11
automatic paging of panels 28
axes
    See also discrete axis
    See also linear axis
    See also log axis
    See also time axis
    about 20, 221
    axis assignment examples 134, 141, 148
    creating axis-aligned statistics table 240
    creating multi-line tick values 239
    creating tick values with Unicode 238
    options supported 221
    SGPLOT procedure 22
AXIS= option, REFLINE statement (SGPLOT) 114

B

background images in graphs 244
band plots
    about 89
    graph examples 61
    grouped 91, 124
    grouped with transparency 91
    overlay 92, 124
    overlay with constant lower limit 92
    overlay with curve labels 93
    overlay with scatter and series plots 122
    overlay with scatter plots 123
    overlay with series plots 123
    overlay with step plots 93
    roles and options supported 89

BAND statement, SGPLOT procedure
  CURVELABELLOWER= option 93
  CURVELABELPOS= option 93
  CURVELABELUPPER= option 93
  LEGENDLABEL= option 92
  LOWER= option 92
  MODELNAME= option 90, 93
  TRANSPARENCY= option 124
bar charts
  See also horizontal bar charts
  See also vertical bar charts
  bar-bar overlay 49
  bar-line overlay 49, 218
  classification panel example 55
  decorative skins and 15
  discrete order for 229
  graph examples 47
  grouped 265
  grouped with skins 48, 51
  HighLow 110–111
  multiple with patterns, fill colors, skins 51
  parametric 100–104
  principles of effective graphics and 7–8
  SGPLOT procedure 22
  stacked with skins 48
  with multiple responses 50
  with patterns 50
BARWIDTH= option
  VBAR statement (SGPLOT) 50–51
  VBARPARM statement (SGPLOT) 125
BASELINE= option, NEEDLE statement (SGPLOT) 96
Berra, Yogi 310
BINSTART= option, HISTOGRAM statement (SGPLOT) 164–165
BINWIDTH= option, HISTOGRAM statement (SGPLOT) 164–165
Bonaparte, Napoleon 220
bond maturity and financial trend graph 311, 313
BORDER option, ODS GRAPHICS statement 328
box plots
  grouped 42
  horizontal 174–176
  interval 43
  row lattice example 55
  SGPLOT procedure 22
  vertical 42, 161, 169–173
BOXWIDTH= option, VBOX statement (SGPLOT) 171
BRADIUSMAX= option, BUBBLE statement (SGPLOT) 54, 248
BRADIUSMIN= option, BUBBLE statement (SGPLOT) 248
BREAK option
  HLINE statement (SGPLOT) 207
  SERIES statement (SGPLOT) 82
  STEP statement (SGPLOT) 88
  VLINE statement (SGPLOT) 200
bubble legends 248
bubble plots
  about 105
  graph examples 54, 61, 106
  grouped 106–107
  roles and options supported 105
  with negative response data 107
BUBBLE statement, SGPLOT procedure
  BRADIUSMAX= option 54, 248
  BRADIUSMIN= option 248
  NOFILL option 54
bullet charts 190, 197
business graphs
  about 311
  danger of high P/E ratios 314
  financial trend and bond maturity graph 311, 313
  oil consumption trend by country 315
  product sales and target graph 311, 316
  social network 311, 317
  stock price and volume chart 311–312

C

Carr, Daniel 4
categorization plots
    See also bar charts
    See also dot plots
    See also line charts
    about 181
    combining statements 29
    roles and options supported 182
    SGPANEL procedure 181, 258
    SGPLOT procedure 22, 62, 181
CATEGORYORDER= option
    discrete order and 229
    DOT statement (SGPLOT) 216
    HBAR statement (SGPLOT) 193
    VBAR statement (SGPLOT) 187
    VLINE statement (SGPLOT) 209
cell graphs 20, 307
classification panels
    See also SGPANEL procedure
    about 20–21, 37, 257–258
    automatic paging 28
    bar chart panels 55
    box plot row lattices 55
    column lattice 265
    grouped bar chart 265
    histogram lattice 56
    lattice by origin and type 262
    mileage by horsepower and type 261
    mileage by type 261
    paging of large panels 266
    panel by origin and type 262
    panel by type 263–264
    row lattice by type 263–264
    scatter panels 56
Cleveland, William S. 3
CLI option
    PBSPLINE statement (SGPLOT) 147–148
    REG statement (SGPLOT) 40, 133–134
CLIP option, ELLIPSE statement (SGPLOT) 155

CLM option
    LOESS statement (SGPLOT) 140–141, 157
    PBSPLINE statement (SGPLOT) 146–147
    REG statement (SGPLOT) 40, 132
CLMTRANSPARENCY= option, REG statement (SGPLOT) 40
CLOSE= option, HIGHLOW statement (SGPLOT) 53, 109
CLUSTERWIDTH= option
    HBOX statement (SGPLOT) 176
    NEEDLE statement (SGPLOT) 96
    SERIES statement (SGPLOT) 83
COLAXIS statement, SGPANEL procedure
    about 25–26, 33, 257, 259
    box plot row lattice example 55
    histogram lattice example 56
color graphs 14–15
COLUMN= option, PANELBY statement (SGPANEL) 263
COLUMNLATTICE layout (SGPANEL procedure) 28
comparative and matrix graphs 37, 57
comparative and matrix plots
    See also SGSCATTER procedure
    about 269
    comparative graph 277, 279
    grouped comparative graph 278
    grouped comparative plot 278
    grouped scatter plots 284
    heat map 280
    multiple plot requests 274
    plot grids 272
    plot with fit 273
    plot with fit and ellipse 273
    rectangular matrix 280
    scatter plot matrix 282, 284
    scatter plot matrix with diagonals and ellipse 283
    scatter plot matrix with histograms 283
    scatter plots 271
    scatter plots with attributes 271

comparative and matrix plots (*continued*)
  uniform side-by-side plots 275
COMPARE statement, SGSCATTER
    procedure
  about 31, 269, 276
  comparative graph 277, 279
  grouped comparative graph 278
  grouped comparative plot 278
  heat map 280
  options supported 276
  rectangular matrix 280
confidence plots
  See fit and confidence plots
CONNECT= option, VBOX statement
    (SGPLOT) 173
Creating More Effective Graphs (Robbins)
    3
curve labels
  horizontal line charts with 210
  images as 245–246
  Loess plots and 144
  overlay band plots with 93
  overlay step plots with 88
  penalized b-spline plots and 151
  regression plots and 137
  vertical line charts with 203
CURVELABEL= option
  HLINE statement (SGPLOT) 210
  LINEPARM statement (SGPLOT) 116
  LOESS statement (SGPLOT) 144
  PBSPLINE statement (SGPLOT) 151
  REG statement (SGPLOT) 137
  SERIES statement (SGPLOT) 80, 123
  STEP statement (SGPLOT) 88
  VLINE statement (SGPLOT) 203
CURVELABELLOC= option, LINEPARM
    statement (SGPLOT) 116
CURVELABELLOWER option, BAND
    statement (SGPLOT) 93
CURVELABELPOS= option
  BAND statement (SGPLOT) 93
  LINEPARM statement (SGPLOT) 116
  SERIES statement (SGPLOT) 123

CURVELABELUPPER option, BAND
    statement (SGPLOT) 93

D

data labels
  dot plots with 214
  horizontal bar charts with 193
  horizontal line charts with 207
  parametric bar charts with 101–102
  scatter plots with 74–77, 135, 142, 149
  series plots with 81
  vertical bar charts with 185
  vertical box plots with 173
  vertical line charts with 200, 202
  waterfall charts with 119
DATALABEL option
  DOT statement (SGPLOT) 214
  HBAR statement (SGPLOT) 193
  HLINE statement (SGPLOT) 207
  LOESS statement (SGPLOT) 142
  PBSPLINE statement (SGPLOT) 149
  REG statement (SGPLOT) 135
  SCATTER statement (SGPLOT)
      74–75, 156
  SERIES statement (SGPLOT) 81
  VBAR statement (SGPLOT) 47, 185
  VBARPARM statement (SGPLOT) 102
  VBOX statement (SGPLOT) 173
  VLINE statement (SGPLOT) 200
DATALABELATTRS= option
  DOT statement (SGPLOT) 214
  HBAR statement (SGPLOT) 193
  HLINE statement (SGPLOT) 207
  VBAR statement (SGPLOT) 185
  VLINE statement (SGPLOT) 200
DATALABELPOS= option
  VBAR statement (SGPLOT) 187
  VBARPARM statement (SGPLOT) 101
  VLINE statement (SGPLOT) 202
DATASKIN= option
  HBAR statement (SGPLOT) 192
  VBAR statement (SGPLOT) 50–51,
      184

VBARPARM statement (SGPLOT) 102
decorative skins, effective graphics and 15
DEGREE= option
 PBSPLINE statement (SGPLOT) 149
 REG statement (SGPLOT) 40, 135
density plots
 about 166
 density curve options 167
 graphic example 161
 histograms with 177
 normal density curve 167
 roles and options supported 166
DENSITY statement
 SGPANEL procedure 25, 56, 257
 SGPLOT procedure 23, 166–168
Descartes, Rene 332
device-based graphics 33
discrete axis
 about 221
 discrete order 229
 fit policy 229
 reference line on 114
 scatter plots with clustered groups on 72
 scatter plots with groups on 72
DISCRETEMAX= option, ODS GRAPHICS statement 328–329
DISCRETEOFFSET= option
 DOT statement (SGPLOT) 217
 HBAR statement (SGPLOT) 197
 HLINE statement (SGPLOT) 211
 REFLINE statement (SGPLOT) 77, 114, 294
 SCATTER statement (SGPLOT) 73
 SERIES statement (SGPLOT) 82, 125
 VBAR statement (SGPLOT) 50–51, 190
 VBARPARM statement (SGPLOT) 102
 VBOX statement (SGPLOT) 171
 VLINE statement (SGPLOT) 204
DISCRETEORDER= option, XAXIS statement (SGPLOT) 229

DISPLAY= option
 X2AXIS statement (SGPLOT) 77
 YAXIS statement (SGPLOT) 52
distribution plots
 See also box plots
 See also density plots
 See also histograms
 about 161
 combining 29, 177
 SGPANEL procedure 161, 258
 SGPLOT procedure 22, 62, 161
documents, small graphs in 333–334
dot plots
 about 212
 combining 218
 discrete order for 229
 graph examples 181, 213
 grouped 216
 HLINE overlay 218
 overlaid 217
 overlaid with discrete offsets 217
 roles and options supported 212
 SGPLOT procedure 22
 with confidence limits 215
 with data labels 214
 with limits 46
 with marker attributes 213
 with reference line 214
 with response sorting 216
 with upper limits 215
DOT statement, SGPLOT procedure
 about 212
 ALPHA= option 215
 CATEGORYORDER= option 216
 DATALABEL option 214
 DATALABELATTRS= option 214
 DISCRETEOFFSET= option 217
 GROUP= option 216
 GROUPDISPLAY= option 212
 LEGENDLABEL= option 217
 LIMITS= option 215
 MARKERATTRS= option 213
 NUMSTD= option 215

DOWN= option, KEYLEGEND statement
    (SGPLOT) 231
DPI
    graph size and 336
    output destination and 327

E

Edison, Thomas Alva 256
effective graphics
    decorative skins and 15
    principles of 3–9
Einstein, Albert 60, 160, 234, 326
ellipse plots
    about 152
    attribute control 155
    clipping 154
    filled 153
    graph examples 129, 153, 273
    overlaid 156
    overlaid filled 156
    roles and options supported 152
    scatter plot matrix with 283
    setting type 155
    with confidence level 154
ELLIPSE statement, SGPLOT procedure
    about 152
    ALPHA= option 153–154
    CLIP option 154
    FILL option 153, 156
    FILLATTRS= option 155
    LINEATTRS= option 155
    OUTLINE option 153
    TYPE= option 155
ERRORBARATTRS= option, STEP
    statement (SGPLOT) 45
eye irritation, distribution of
    using SGPANEL procedure 303
    using SGPLOT procedure 304

F

FILL option, ELLIPSE statement (SGPLOT)
    153, 156

FILLATTRS= option, ELLIPSE statement
    (SGPLOT) 155
financial trend and bond maturity graph
    311, 313
fit and confidence plots
    See also ellipse plots
    See also Loess plots
    See also penalized b-spline plots
    See also regression plots
    about 129
    combining 29, 157
    comparing fits 157
    comparing fits with CLM 157
    fit plots 39
    fit plots with confidence limits 40
    fit plots with transparent markers 39
    grouped fit plot with limits 40
    roles and options 130
    SGPANEL procedure 129, 258
    SGPLOT procedure 22, 62, 129
FITPOLICY= option, XAXIS statement
    (SGPLOT) 119, 229
floating bars and bands 125
FOOTNOTE statement 33, 245
forest plots 241–242, 288–289
FREQ= option
    HISTOGRAM statement (SGPLOT)
        163
    SCATTER statement (SGPLOT) 81

G

Gates, W.I.E. 2
GCHART procedure 33
GOPTIONS global statement 33
GPLOT procedure 33
graph output
    about 333
    combining graph size and DPI 336
    creating large graphs for presentations
        334–335
    creating small graphs for documents
        333–334

impact of graph size on system
    resources 336
Graph Template Language (GTL) 11
graphs
    See also business graphs
    See also classification panels
    See also health and life sciences graphs
    See also single-cell graphs
    about 20, 37
    automatically producing from
        procedures 10–11
    background images in 244
    color 14–15
    comparative and matrix 37, 57
    decorative skins and 15
    Graph Template Language 11
    gray-scale 14–15
    principles of effective graphics 3–9
    statistical graphics procedures and
        11–12
gray-scale graphs 14–15
GRID option
    PLOT statement (SGPANEL) 275
    X2AXIS statement (SGPLOT) 134,
        141, 148
    Y2AXIS statement (SGPLOT) 134,
        141, 148
GROUP= option
    DOT statement (SGPLOT) 216
    HBAR statement (SGPLOT) 195
    HBOX statement (SGPLOT) 176
    HLINE statement (SGPLOT) 210
    LOESS statement (SGPLOT) 143
    PBSPLINE statement (SGPLOT) 150
    REG statement (SGPLOT) 136
    SCATTER statement (SGPLOT) 65,
        72, 76
    style elements and 323
    VBAR statement (SGPLOT) 51, 188
GROUPDISPLAY= option
    DOT statement (SGPLOT) 212
    HBAR statement (SGPLOT) 191, 195
    HLINE statement (SGPLOT) 205

SCATTER statement (SGPLOT) 72
SERIES statement (SGPLOT) 83
VBAR statement (SGPLOT) 48, 51,
    188
VBOX statement (SGPLOT) 170
GROUPMAX= option, ODS GRAPHICS
    statement 328–329
GSEG output format 33
GTL (Graph Template Language) 11

H

HBAR statement, SGPLOT procedure
    about 191
    ALPHA= option 194
    CATEGORYORDER= option 193
    DATALABEL option 193
    DATALABELATTRS= option 193
    DATASKIN= option 192
    DISCRETEOFFSET= option 197
    GROUP= option 195
    GROUPDISPLAY= option 191, 195
    LIMITS= option 52, 194
    NOFILL option 192
    NOOUTLINE option 192
    NUMSTED= option 194
    STAT= option 52
HBARPARM statement, SGPLOT procedure
    about 100
    LIMITATTRS= option 104
    with statistics 126
HBOX statement, SGPLOT procedure
    about 174
    CLUSTERWIDTH= option 176
    GROUP= option 176
    LINEATTRS= option 176
    NOTCHES option 175
health and life sciences graphs
    about 287
    adverse event timelines 291–292
    concomitant medications 306
    creating 2 x 2 cell graph using SGPLOT
        procedure 307

health and life sciences graphs (*continued*)
  distribution of eye irritation using
    SGPANEL procedure   303
  distribution of eye irritation using
    SGPLOT procedure   304
  forest plots   288–289
  immunology profile by treatment   299
  LFT patient profile   301
  LFT safety panel, baseline vs. study
    298
  maximum LFT values by treatment   293
  median of lipid profile over time by
    treatment   294
  most frequent on-therapy adverse
    events by frequency   300
  panel of LFT values   302
  QTc change from baseline over time by
    treatment   295
  QTc change graph with annotated at
    risk values   296
  survival plots   290
  vital signs by time point name   305
heat maps   280
HEIGHT= option, ODS GRAPHICS
  statement   15, 328
HIGH= option, HIGHLOW statement
  (SGPLOT)   109
HIGHCAP option, HIGHLOW statement
  (SGPLOT)   53
HIGHLABEL= option, HIGHLOW statement
  (SGPLOT)   110
HighLow bar charts   110–111
HighLow plots
  about   108
  bar chart examples   110–111
  graph examples   53, 61, 109
  roles and options supported   108
  vertical vector plots and   99
  with end caps   110
  with open and close values   109
HIGHLOW statement, SGPLOT procedure
  about   108
  CLOSE= option   53, 109

HIGH= option   109
HIGHCAP option   53
HIGHLABEL= option   110
LOW= option   109
LOWCAP option   53
LOWLABEL= option   110–111, 292
OPEN= option   53, 109
TYPE= option   53, 109
HISTOGRAM statement
  SGPANEL procedure   25, 56, 257
  SGPLOT procedure   23, 41, 162–165
histograms
  about   162
  classification panel example   56
  graph examples   41, 161, 163
  multiple   165
  roles and options supported   162
  scatter plot matrix with   283
  SGPLOT procedure   22
  with BINSTART   164
  with BINWIDTH   164
  with density plots   41
  with frequency   163
  with NBINS   164
  with normal density plots   177
HLINE statement, SGPLOT procedure
  about   205
  ALPHA= option   208
  BREAK option   207
  CURVELABEL= option   210
  DATALABEL option   207
  DATALABELATTRS= option   207
  DISCRETEOFFSET= option   211
  dot plot overlay   218
  GROUP= option   210
  LEGENDLABEL= option   211
  LIMITS= option   209
  LINEATTRS= option   206
  MARKERATTRS= option   206
  MARKERS option   206
  NUMSTD= option   209
horizontal bar charts
  about   191

bullet charts   197
  graph examples   52, 181
  grouped using SGPANEL procedure   196
  overlaid with discrete offset   197
  roles and options supported   191
  with adjacent groups   195
  with confidence limits   194
  with data labels   193
  with fill attributes and data skins   192
  with groups   195
  with groups and pattern fills   196
  with limits   52
  with no fill   192
  with reference line   193
  with upper limit   194
horizontal box plots
  about   174
  graph examples   161, 175
  grouped   176
  grouped with CLUSTERWIDTH   176
  roles and options supported   174
  with notches   175
horizontal line charts
  about   205
  graph examples   181, 206
  grouped   210
  overlaid   211
  overlaid with discrete offset   211
  roles and options supported   205
  with break   207
  with confidence limits   208
  with curve labels   210
  with data labels   207
  with markers   206
  with reference line   208
  with response sorting   209
  with upper limits   209

I

IMAGE function   236
IMAGEMAP option, ODS GRAPHICS statement   328–329
IMAGENAME= option, ODS GRAPHICS statement   328–329
images as curve labels   245–246
immunology profile by treatment   299
Information Visualization (Carr)   4
INSET statement, SGPLOT procedure   24, 85, 106, 222
insets
  about   222
  options supported   222
  rich-text   232
  SGPLOT procedure   22
  types of   232
INTEGER option, YAXIS statement (SGPLOT)   224
INTERPOLATION= option, LOESS statement (SGPLOT)   142
interval box plots   43
INTERVAL= option, XAXIS statement (SGPLOT)   227

J

JOIN option, PLOT statement (SGPANEL)   275
JPG output format   33
JUSTIFY= option, STEP statement (SGPLOT)   86

K

Kay, Alan   18
KEYLEGEND statement, SGPANEL procedure   33, 221, 259
KEYLEGEND statement, SGPLOT procedure
  about   23–24, 33, 221
  ACROSS= option   231
  bubble plot examples   106
  custom legends   230
  DOWN= option   231
  fit plot examples   40
  LOCATION= option   231
  POSITION= option   231
  scatter plot examples   38, 65–66, 121

KEYLEGEND statement, SGPLOT procedure (*continued*)
   series plot examples 79–80
   step plot examples 85
   VALUEATTRS= option 122

L

LABEL= option
   XAXIS statement (SGPLOT) 206, 213
   YAXIS statement (SGPLOT) 199
LABELMAX= option, ODS GRAPHICS statement 328–329
LABELPOS= option, REFLINE statement (SGPLOT) 113
LATTICE layout (SGPANEL procedure) 27
LAYOUT= option, PANELBY statement (SGPANEL)
   about 257
   box plot row lattice 55
   grouped bar chart 189, 196
   lattice by origin and type 262
   paging of large panels 266
   row lattice by type 263
LEGEND global statement 33
LEGEND option, SGSCATTER procedure 33, 222
LEGENDLABEL= option
   BAND statement (SGPLOT) 92
   custom legends and 230
   DOT statement (SGPLOT) 217
   HLINE statement (SGPLOT) 211
   PBSPLINE statement (SGPLOT) 151
   REG statement (SGPLOT) 137
   SCATTER statement (SGPLOT) 71
   VLINE statement (SGPLOT) 204
legends
   about 20, 221–222
   ACROSS option 231
   automatic 230
   bubble 248
   custom 230
   disabling 133, 140, 147
   DOWN option 231
   grouped series plots with 79
   LOCATION option 231
   options supported 222
   POSITION option 231
   SGPLOT procedure 22
LFT patient profile 301
LFT safety panel, baseline vs. study 298
LFT values, panel of 302
life sciences graphs
   See health and life sciences graphs
LIFETEST procedure 290
LIMITATTRS= option, HBARPARM statement (SGPLOT) 104
LIMITS= option
   DOT statement (SGPLOT) 215
   HBAR statement (SGPLOT) 52, 194
   HLINE statement (SGPLOT) 209
   VBAR statement (SGPLOT) 186
   VLINE statement (SGPLOT) 202
line charts
   bar-line overlay 49, 218
   discrete order for 229
   horizontal 181
   principles of effective graphics and 9
   vertical 181, 198–204
LINE function 236, 248
line plots, parametric 115–116
linear axis
   about 221
   specifying MIN/MAX 223
   specifying tick lists 223
   TICKVALUEFORMAT option 228
   VALUESHINT option 224
LINEATTRS= option
   ELLIPSE statement (SGPLOT) 155
   HBOX statement (SGPLOT) 176
   HLINE statement (SGPLOT) 206
   NEEDLE statement (SGPLOT) 95
   REG statement (SGPLOT) 39
   STEP statement (SGPLOT) 45, 86
   VECTOR statement (SGPLOT) 98
   VLINE statement (SGPLOT) 49, 199
LineParm plots 61

LINEPARM statement, SGPLOT procedure
    CURVELABEL= option   116
    CURVELABELLOC= option   116
    CURVELABELPOS= option   116
lipid profile, median of, over time by treatment   294
LOCATION= option, KEYLEGEND statement (SGPLOT)   231
Loess plots
    about   138
    attribute control   141
    automatic legend   140
    axis assignment   141
    basic fit   139
    CLM band   140
    cubic Loess fit   142
    curve labels   144
    fit plot with smooth curves   139
    graph examples   129
    grouped fits   143
    overlaid fits   144
    roles and options supported   138
    scatter data labels   142
    SGPLOT procedure   22
    single fit with a group   143
LOESS statement, SGPLOT procedure
    ALPHA= option   140
    CLM option   140–141, 157
    CURVELABEL= option   144
    DATALABEL option   142
    GROUP= option   143
    INTERPOLATION= option   142
    NOLEGFIT option   140
    NOMARKERS option   139, 143–144
    SMOOTH= option   139
log axis
    about   221
    LINEAR style   226
    log expand style   225
    log exponent style   225
    MINOR option   228
    specifying axis MIN/MAX   223

LOGBASE= option, YAXIS statement (SGPLOT)   226
LOGSTYLE= option, YAXIS statement (SGPLOT)   226
LOW= option, HIGHLOW statement (SGPLOT)   109
LOWCAP option, HIGHLOW statement (SGPLOT)   53
LOWER= option, BAND statement (SGPLOT)   92
LOWLABEL= option, HIGHLOW statement (SGPLOT)   110–111, 292

M

magnitude perception, accuracy of   6
MARKERATTRS= option
    DOT statement (SGPLOT)   213
    HLINE statement (SGPLOT)   206
    SCATTER statement (SGPLOT)   66–67
    VLINE statement (SGPLOT)   199
MARKERCHAR= option, SCATTER statement (SGPLOT)
    about   74
    bubble plot with negative response data   107
    forest plots   288
    HBARPARM with statistics   126
    median of lipid profile over time by treatment   294
    scatter with axis-aligned statistics table   77
    scatter with marker characters   77
    VBARPARM with statistics   126
MARKERS option, HLINE statement (SGPLOT)   206
matrix graphs   37, 57
matrix plots
    See comparative and matrix plots
MATRIX statement, SGSCATTER procedure
    about   31, 269, 281
    grouped scatter plots   284
    options supported   281

MATRIX statement, SGSCATTER procedure (*continued*)
   scatter plot matrix   282, 284
   scatter plot matrix with diagonals and ellipse   283
   scatter plot matrix with histograms   283
   START= option   284
MAX= option, XAXIS statement (SGPLOT)   223
MAXLEGENDAREA= option, ODS GRAPHICS statement   328–329
medications, concomitant   306
MIN= option
   XAXIS statement (SGPLOT)   223
   YAXIS statement (SGPLOT)   79, 85
MINOR option
   XAXIS statement (SGPLOT)   228
   YAXIS statement (SGPLOT)   225
MODELNAME= option, BAND statement (SGPLOT)   90, 93

N

needle plots
   about   94
   graph examples   61, 95
   grouped   95
   roles and options supported   94
   with baseline   96
   with discrete groups   96
NEEDLE statement, SGPLOT procedure
   BASELINE= option   96
   CLUSTERWIDTH= option   96
   LINEATTRS= option   95
Nietzsche, Friedrich   180
NOARROWHEADS option, NEEDLE statement (SGPLOT)   99
NOAUTOLEGEND option, SGPLOT procedure   68, 107, 133, 140, 147
NOBORDER option, PANELBY statement (SGPANEL)   189, 196
NOCYCLEATTRS option, SGPLOT procedure   68
NOFILL option
   BUBBLE statement (SGPLOT)   54
   HBAR statement (SGPLOT)   192
   HISTOGRAM statement (SGPLOT)   164–165
   VBAR statement (SGPLOT)   184
   VBOX statement (SGPLOT)   171
NOLEGFIT option
   LOESS statement (SGPLOT)   140
   PBSPLINE statement (SGPLOT)   147
   REG statement (SGPLOT)   133
NOMARKERS option
   LOESS statement (SGPLOT)   139, 143–144
   PBSPLINE statement (SGPLOT)   146, 150–151
   REG statement (SGPLOT)   132, 136–137
NOTCHES option
   HBOX statement (SGPLOT)   175
   VBOX statement (SGPLOT)   173
NOTIMESPLIT option, XAXIS statement (SGPLOT)   227
NOUTLINE option
   HBAR statement (SGPLOT)   192
   VBAR statement (SGPLOT)   184
NOVARNAME option, PANELBY statement (SGPANEL)   261
NUMSTD= option
   DOT statement (SGPLOT)   215
   HBAR statement (SGPLOT)   194
   HLINE statement (SGPLOT)   209
   VBAR statement (SGPLOT)   186
   VLINE statement (SGPLOT)   202

O

ODS destination options   32, 321, 327
ODS ESCAPECHAR option   232, 238
ODS GRAPHICS statement
   ANTIALIAS option   328
   ANTIALIASMAX= option   44, 79, 328
   BORDER option   328
   DISCRETEMAX= option   328–329
   GROUPMAX= option   328–329

HEIGHT= option 15, 328
IMAGEMAP option 328–329
IMAGENAME= option 328–329
LABELMAX= option 328–329
MAXLEGENDAREA= option 328–329
OUTPUTFMT= option 328–329
PANELCELLMAX= option 328–329
RESET option 328–329
SCALE option 328–329
TIPMAX= option 328, 330
WIDTH= option 15, 328, 330
ODS Graphics system
  annotation and 236
  automatically producing graphs 10–11
  options supported 328–330
  SG procedures and 19
  template-based graphs and 33
ODS SELECT statement 10
OFFSETMAX= option
  annotations and 237
  YAXIS statement (SGPLOT) 46, 95, 102, 173
OFFSETMIN= option
  annotations and 237
  XAXIS statement (SGPLOT) 71, 114
  YAXIS statement (SGPLOT) 118
oil consumption trend by country 315
ONEPANEL option, PANELBY statement (SGPANEL) 263
OPEN= option, HIGHLOW statement (SGPLOT) 53, 109
OUTLINE option, ELLIPSE statement (SGPLOT) 153
OUTPUTFMT= option, ODS GRAPHICS statement 328
OVAL function 236, 248

P

P/E ratios 314
PAD= option, SGPLOT procedure 238, 240
paging of large panels 266
PANEL layout (SGPANEL procedure) 26–27
PANELBY statement, SGPANEL procedure
  about 24–25, 257, 259
  COLUMN= option 263
  LAYOUT= option 55, 189, 196, 257, 266
  NOBORDER option 189, 196
  NOVARNAME option 261
  ONEPANEL option 263
  roles and options supported 260
  UNISCALE= option 264
PANELCELLMAX= option, ODS GRAPHICS statement 328–329
parametric bar charts
  graph examples 101
  grouped 102
  grouped with data labels 102
  horizontal 104
  horizontal with limits 104
  overlay 103
  overlay with skins and offsets 103
  vertical 101–102
  with limits and data labels 101
parametric line plots 115–116
Pareto chart 125
PATTERN global statement 33
PBSPLINE statement, SGPLOT procedure
  ALPHA= option 146–147
  CLI option 147–148
  CLM option 146–147
  CURVELABEL= option 151
  DATALABEL option 149
  DEGREE= option 149
  GROUP= option 150
  LEGENDLABEL= option 151
  NOLEGFIT option 147
  NOMARKERS option 146, 150–151
PDF output format 33
penalized b-spline plots
  about 145
  attribute control 148
  automatic legend 147
  axis assignments 148
  basic parametric 146

penalized b-spline plots (*continued*)
    changing degree of   149
    CLI band   147
    CLM band   146
    curve labels   151
    graph examples   129, 273
    grouped fits   150
    overlaid fits   151
    roles and options supported   145
    scatter data labels   149
    single fit within a group   150
pie charts   3–4, 6
PLOT statement, SGSCATTER procedure
    about   30, 269–270
    GRID option   275
    JOIN option   275
    options supported   270
    scatter plots   271
    UNISCALE= option   272, 275
plot statements
    See SGPLOT procedure
    See specific statements
plots
    See also reference line
    See also SGPLOT procedure
    See also specific types of plots
    about   61
    combining   29, 120–126
    SGPANEL procedure   258
    waterfall charts   117–119
PNG output format   33
POLYCONT function   235–236
POLYGON function   235–236
POLYLINE function   235–236, 249–250
POSITION= option, KEYLEGEND statement (SGPLOT)   231
presentations, large graphs in   334–335
principles of effective graphics
    about   3–4
    accuracy of magnitude perception   6
    additional information   3
    proximity increases accuracy of comparisons   8
    short-term memory and   9
    simplifying and reducing clutter   8
    usage of 3-D graphs   7
    visual perception   4–5
procedures, automatic graphs from   10–11
product sales and target graph   311, 316
PS output format   33

Q

QTc change from baseline over time by treatment   295
QTc change graph with annotated at risk values   296

R

RECTANGLE function   236, 248
reference line
    about   112
    dot plots with   214
    graph examples   61
    horizontal bar charts with   193
    horizontal line charts with   208
    multiple   114
    on discrete axis   114
    roles and options supported   112
    scatter plots with   113
    SGPANEL procedure   25
    SGPLOT procedure   22
    vertical bar charts with   185
    vertical line charts with   201
REFLINE statement, SGPANEL procedure   259
REFLINE statement, SGPLOT procedure
    about   24, 71
    AXIS= option   114
    bar charts and   185, 193
    DISCRETEOFFSET= option   77, 114, 294
    dot plots and   214
    LABELPOS= option   113
    line charts and   201, 208
    step and scatter plot example   122
REG procedure   10, 134

REG statement, SGPLOT procedure
    ALPHA= option   132–133
    CLI option   40, 133–134
    CLM option   40, 132
    CLMTRANSPARENCY= option   40
    CURVELABEL= option   137
    DATALABEL option   135
    DEGREE= option   40, 135
    GROUP= option   136
    LEGENDLABEL= option   137
    LINEATTRS= option   39
    NOLEGFIT option   133
    NOMARKERS option   132, 136–137
regression plots
    about   131
    attribute control   134
    automatic legend   133
    axis assignment   134
    basic linear fit   132
    CLI band   133
    CLM band   132
    curve labels   137
    graph examples   129
    grouped fits   136
    overlaid fits   137
    polynomial regression fit   135
    roles and options supported   131
    scatter data labels   135
    SGPLOT procedure   22
    single fit with a group   136
RESET option
    GOPTIONS statement   33
    ODS GRAPHICS statement   328–329
REVERSE option
    XAXIS statement (SGPLOT)   197
    YAXIS statement (SGPLOT)   190
Robbins, Naomi   3
ROWAXIS statement, SGPANEL procedure
    about   25–26, 33, 259
    histogram lattice example   56
ROWLATTICE layout (SGPANEL procedure)   28

S

SAS/GRAPH   236
SCALE option
    ODS GRAPHICS statement   328–329
SCALE= option
    DENSITY statement (SGPLOT)   167
scatter plot matrix
    graph examples   57, 282, 284
    with diagonals and ellipse   283
    with histograms   283
scatter plots
    about   64
    classification panel example   56
    comparative   57
    graph examples   38, 61, 65, 271
    grouped   65, 284
    grouped with grid lines   38
    grouped with X and Y limits   69
    overlaid fits   137, 144, 151
    overlay of   73, 121
    overlay with band plots   123
    overlay with discrete offset   73
    overlay with series and band plots   122
    overlay with step plots   122
    roles and options supported   64
    SGPLOT procedure   22
    single fit with a group   136, 143, 150
    two plot overlay   66
    with attributes   271
    with clustered groups on discrete axis   72
    with custom attributes   66
    with data labels   74–77, 135, 142, 149
    with groups on discrete axis   72
    with lower and upper limits for X and Y   68
    with reference line   113
    with reflected ticks on Y2   71
    with X, Y, and Y2 axis   71
    with X2 and Y2 axes   70
    with X lower and upper error limits   68
    with Y2 axis   70

scatter plots (continued)
    with Y lower and upper error limits   67
    with Y upper error limit   67
SCATTER statement, SGPLOT procedure
    about   39
    DATALABEL option   74–75, 156
    DISCRETEOFFSET= option   73
    FREQ= option   81
    GROUP= option   65, 72, 76
    LEGENDLABEL= option   71
    MARKERATTRS= option   66–67
    MARKERCHAR= option   74, 77, 107, 126, 288, 294
    overlaid fits   137, 144, 151
    single fit with a group   136, 143, 150
series plots
    graph examples   43, 61, 79
    grouped   44
    grouped with curved labels   80
    grouped with legends   79
    grouped with markers   80–81
    multi-response   44
    overlay with band plots   123
    overlay with discrete offset   82
    overlay with scatter and band plots   122
    overlay with scatter plots   121
    roles and options supported   78
    SGPLOT procedure   22
    with breaks   82
    with data labels   81
    with group cluster   83
    with group overlay   83
SERIES statement, SGPLOT procedure
    BREAK option   82
    CLUSTERWIDTH= option   83
    CURVELABEL= option   80, 123
    CURVELABELPOS= option   123
    DATALABEL option   81
    DISCRETEOFFSET= option   82, 125
    GROUPDISPLAY= option   83
    Pareto chart   125
SG (Statistical Graphics) procedures
    See also specific procedures

    about   11
    analysis of data   12
    annotation support   237
    automatic procedures   21
    combining statements   29
    key concepts   19–21
    legends and   230
    post-analysis data presentation   12
    pre-analysis data exploration   11
    styles and their usage   32
    template-based vs. device-based graphics   33
SGANNO= option, SGPLOT procedure   235
SGPANEL procedure
    See also classification panels
    See also COLAXIS statement, SGPANEL procedure
    See also PANELBY statement, SGPANEL procedure
    about   19, 24, 258–259
    annotation support   237
    automatic paging   28
    axis support   221
    COLUMNLATTICE layout   28
    DENSITY statement   25, 56, 257
    distribution of eye irritation using   303
    fit and confidence plots   129
    grouped bar charts   189
    HISTOGRAM statement   25, 56, 257
    KEYLEGEND statement   33, 221, 259
    LATTICE layout   27
    PANEL layout   26–27
    plots supporting   61
    REFLINE statement   259
    ROWAXIS statement   25–26, 33, 56, 259
    ROWLATTICE layout   28
SGPLOT procedure
    See also KEYLEGEND statement, SGPLOT procedure
    See also REFLINE statement, SGPLOT procedure

See also SCATTER statement, SGPLOT procedure
See also XAXIS statement, SGPLOT procedure
See also YAXIS statement, SGPLOT procedure
about   19, 22, 62
annotation support   237, 295
axis support   221
BAND statement   90, 92–93, 124
BUBBLE statement   54, 248
creating cell graphs with   307
DENSITY statement   23, 166–168
distribution of eye irritation using   304
DOT statement   212–217
fit and confidence plots   129
HBAR statement   52, 191–197
HBARPARM statement   100, 104, 126
HBOX statement   174–176
HIGHLOW statement   53, 108–111, 292
HISTOGRAM statement   23, 41
HLINE statement   205–211
INSET statement   24, 85, 106
LINEPARM statement   115–116
LOESS statement   139–144, 157
NEEDLE statement   95–96
NOAUTOLEGEND option   68, 107, 133, 140, 147
NOCYCLEATTRS option   68
options supported   62
PAD= option   238, 240
PBSPLINE statement   146–151
plot roles and options   63
plot statements supported   62
plots supporting   61
REG statement   39–40, 132–137
SERIES statement   80–83, 123, 125
SGANNO= option   235
single-cell graphs and   19–20, 37
STEP statement   45, 86, 88
typical use case   22–23
VBAR statement   47–48, 50–51, 183–190
VBARPARM statement   100–103, 125–126
VBOX statement   169–173, 295
VECTOR statement   98–99
VLINE statement   49, 198–204
WHERE statement   43
X2AXIS statement   24, 33, 77
Y2AXIS statement   24, 33
SGRENDER procedure   11
SGSCATTER procedure
    about   19, 30
    annotation support   237
    comparative and matrix graphs and   21, 37
    COMPARE statement   31, 269, 276–280
    controlling axes   33
    LEGEND option   33, 222
    MATRIX statement   31, 269, 281–284
    options supported   269
    PLOT statement   30, 269–275
SHOWBINS option, HISTOGRAM statement (SGPLOT)   41
SIGMA= option, DENSITY statement (SGPLOT)   168
single-cell graphs
    about   37
    bar-bar overlay   49
    bar charts   47
    bar charts with multiple responses   50
    bar charts with patterns   50
    bar-line overlay   49
    basic scatter plot   38
    bubble plots   54
    dot plots with limits   46
    fit plots   39
    fit plots with confidence limits   40
    fit plots with transparent markers   39
    grouped bar charts with skins   48, 51
    grouped box plots   42
    grouped fit plot with limits   40

single-cell graphs (*continued*)
  grouped scatter plot with grid lines 38
  grouped series plots 44
  grouped step plots 45
  HighLow plots 53
  histograms 41
  histograms with density plots 41
  horizontal bar charts 52
  horizontal bar charts with limits 52
  interval box plots 43
  multi-response series plots 44
  multiple bar charts with patterns, fill colors, skins 51
  series plots 43
  SGPLOT procedure and 19–20, 37
  stacked bar charts with skins 48
  step plots with limits 45
  vector plots 46
  vertical box plots 42
SMOOTH= option, LOESS statement (SGPLOT) 139
social network (business graphs) 311, 317
START= option, MATRIX statement (SCSCATTER) 284
STAT= option, HBAR statement (SGPLOT) 52
Statistical Graphics procedures
  See SG (Statistical Graphics) procedures
step plots
  about 84
  graph examples 61, 85
  grouped 45, 85
  grouped with center justified markers 86
  grouped with limits and error bar attributes 87
  grouped with markers 86
  grouped with upper limit 87
  overlay with band plots 93
  overlay with break 88
  overlay with curve labels 88
  overlay with scatter plots 122

  roles and options supported 84
  with limits 45
STEP statement, SGPLOT procedure
  BREAK option 88
  CURVELABEL= option 88
  ERRORBARATTRS= option 45
  JUSTIFY= option 86
  LINEATTRS= option 45, 86
Steven's power law 6
stock price and volume chart 311–312
style elements
  about 321–323
  usage considerations 324
  usage precedence 324
STYLE= option (ODS destination) 32, 321
styles
  about 321
  appearance options 32
  custom 32
  pre-defined 32
  SG procedures and 32
survival plots 290
SYMBOL global statement 33
system resources, impact of graph size on 336

T

template-based graphics 33
TEMPLATE procedure 322
TEXT function 236, 248
TEXTCONT function 236
3-D graphs, usage considerations 7
tick lists, specifying 223
TICKVALUEFORMAT= option, XAXIS statement (SGPLOT) 228
time axis
  about 221
  INTERVAL option 227
  MINOR option 228
  NOTIMESPLIT option 227
  specifying axis MIN/MAX 223
  specifying tick lists 223
  TICKVALUEFORMAT option 228

VALUESHINT option 224
TIPMAX= option, ODS GRAPHICS
    statement 328, 330
TITLE statement 33
TRANSPARENCY= option, BAND
    statement (SGPLOT) 124
treatment
    immunology profile by 299
    maximum LFT values by 293
    median of lipid profile over time by 294
    QTc change from baseline over time by 295
Tufte, Edward 3, 8
TYPE= option
    DENSITY statement (SGPLOT) 168
    ELLIPSE statement (SGPLOT) 155
    HIGHLOW statement (SGPLOT) 53, 109
    XAXIS statement (SGPLOT) 43, 172

U

Unicode characters 232, 238
UNISCALE= option
    PANELBY statement (SGPANEL) 264
    PLOT statement (SGPANEL) 272, 275

V

VALUEATTRS= option
    KEYLEGEND statement (SGPLOT) 122
    XAXIS statement (SGPLOT) 125
    YAXIS statement (SGPLOT) 125
VALUES= option, XAXIS statement (SGPLOT) 223
VALUESHINT option, XAXIS statement (SGPLOT) 224
VBAR statement, SGPLOT procedure
    about 183
    ALPHA= option 186
    BARWIDTH= option 50–51
    CATEGORYORDER= option 187
    DATALABEL option 47, 185
    DATALABELATTRS= option 185
    DATALABELPOS= option 187
    DATASKIN= option 50–51, 184
    DISCRETEOFFSET= option 50–51, 190
    GROUP= option 51, 188
    GROUPDISPLAY= option 48, 51, 188
    LIMITS= option 186
    NOFILL option 184
    NOUTLINE option 184
    NUMSTD= option 186
VBARPARM statement, SGPLOT procedure
    about 100
    BARWIDTH= option 125
    DATALABEL option 102
    DATALABELPOS= option 101
    DATASKIN= option 102
    DISCRETEOFFSET= option 103
    Pareto chart 125
    with statistics 126
VBOX statement, SGPLOT procedure
    about 169
    BOXWIDTH= option 171
    CONNECT= option 173
    DATALABEL option 173
    DISCRETEOFFSET= option 171
    GROUPDISPLAY= option 170
    NOFILL option 171
    NOTCHES option 173
    QTc change from baseline over time by treatment 295
vector plots 46–47, 61, 97–99
VECTOR statement, SGPLOT procedure
    ARROWHEADSHAPE= option 99
    LINEATTRS= option 98
    NOARROWHEADS option 99
    XORIGIN= option 98
    YORIGIN= option 98
vertical bar charts
    bar-bar overlay 49
    bar-line overlay 49
    bullet charts 190
    graph examples 47, 181

vertical bar charts (*continued*)
    grouped using SGPANEL procedure   189
    grouped with data labels   102
    grouped with skins   48, 51
    multiple with patterns, fill colors, skins   51
    overlaid with discrete offset   190
    overlay   103
    overlay with skins and offsets   103
    roles and options supported   183
    stacked with skins   48
    with adjacent groups   188
    with an upper limit   186
    with confidence limits   186
    with data labels   185
    with fill attributes and data skins   184
    with groups and pattern fills   189
    with limits and label positioning   187
    with no fill   184
    with patterns   50
    with reference line   185
    with response sorting   187
    with stacked groups   188
vertical box plots
    about   169
    graph examples   42, 161, 170
    grouped   170
    grouped unfilled   171
    on linear axis   172
    overlay   171
    overlay for linear data   172
    roles and options supported   169
    with labels   173
    with notches   173
vertical line charts
    about   198
    graph examples   181, 199
    grouped   203
    overlaid   204
    overlaid with discrete offset   204
    roles and options supported   198
    with break   200
    with CLM and data label position   202
    with confidence limits   201
    with curve labels   203
    with data labels   200
    with markers   199
    with reference line   201
    with upper limits   202
vertical vector plots   99
The Visual Display of Quantitative Information (Tufte)   3
visual perception, effective graphics and   4–5
Visualizing Data (Cleveland)   3
vital signs by time point name   305
VLINE statement, SGPLOT procedure
    about   198
    ALPHA= option   201
    BREAK option   200
    CATEGORYORDER= option   209
    CURVELABEL= option   203
    DATALABEL option   200
    DATALABELATTRS= option   200
    DATALABELPOS= option   202
    DISCRETEOFFSET= option   204
    LEGENDLABEL= option   204
    LIMITS= option   202
    LINEATTRS= option   49, 199
    MARKERATTRS= option   199
    NUMSTD= option   202

W

waterfall charts
    about   117
    graph examples   118
    grouped   119
    roles and options supported   117
    with data labels   119
    with initial value   118
WATERFALL statement, SGPLOT procedure   117
WHERE statement, SGPLOT procedure   43
WIDTH= option, ODS GRAPHICS statement   15, 328, 330

X

XAXIS statement, SGPLOT procedure
   about   23–24, 33
   DISCRETEORDER= option   229
   FITPOLICY= option   119, 229
   INTERVAL= option   227
   LABEL= option   206, 213
   MAX= option   223
   MIN= option   223
   MINOR option   228
   NOTIMESPLIT option   227
   OFFSETMIN= option   71, 114
   REVERSE option   197
   TICKVALUEFORMAT= option   228
   TYPE= option   43, 172
   VALUEATTRS= option   125
   VALUES= option   223
   VALUESHINT option   224
X2AXIS statement, SGPLOT procedure
   about   24, 33
   DISPLAY option   77
   GRID option   134, 141, 148
XORIGIN= option, NEEDLE statement
      (SGPLOT)   98

Y

YAXIS statement, SGPLOT procedure
   about   24, 33
   DISPLAY= option   52
   INTEGER option   224
   LABEL= option   199
   LOGBASE= option   226
   LOGSTYLE= option   226
   MIN= option   79, 85
   MINOR option   225
   OFFSETMAX= option   46, 95, 102, 173
   OFFSETMIN= option   118
   REVERSE option   190
   VALUEATTRS= option   125
Y2AXIS statement, SGPLOT procedure
   about   24, 33
   GRID option   134, 141, 148

Yoda   128
YORIGIN= option, NEEDLE statement
      (SGPLOT)   98

CPSIA information can be obtained at www.ICGtesting.com
Printed in the USA
LVOW09s0205300415

436645LV00003B/19/P